Long Memory Time Series Analysis

Long Memory Time Series Analysis is a comprehensive text which covers long memory time series with the different long memory time series discussed. The authors cover modelling and forecasting using various time series, deploying traditional and machine learning methodologies. The reader also learns recent research trends, such as state space modelling of generalized long memory time series and the use of the tsfGRNN machine learning tool in R. The book starts from autoregressive (AR) and moving average (MA) processes to descriptions of the autoregressive integrated moving average (ARMA) time series, the ARIMA model, and the autoregressive fractionally integrated moving average (ARFIMA) process. The differences of short, intermediate, and long memory processes are highlighted. The reader will gain knowledge of elementary time series through this extensive coverage.

The book discusses generalized Gegenbauer autoregressive moving averages (GARMA) and seasonal GARMA long memory time series and state space modelling of generalized and seasonal GARMA. The extensions of the short and long memory models driven by generalised autoregressive conditionally heteroskedastic (GARCH) errors are also presented. The extensive range of problems linked with generalized Gegenbauer long memory time series are presented to reinforce the reader's conceptual learning. Coverage on the use of time series with high frequency data captured through the latest technological innovations is an invaluable resource to the reader. This learning is done through examples of time series application case studies in medicine, biology, and finance.

The core audience is students attending advanced studies in time series. The book can also be used by researchers and data scientists involved in utilizing time series analysis in a modern context.

Gnanadarsha Sanjaya Dissanayake earned a PhD in statistics, with an emphasis on time series econometrics, at the School of Mathematics and Statistics, University of Sydney, Australia. He is the Senior Biostatistician, New South Wales Ministry of Health, and an Honorary Research Associate, School of Mathematics and Statistics, University of Sydney, Australia.

Hassan Doosti is the Program Director in the Master of Data Science program and the Senior Lecturer in Statistics, School of Mathematical and Physical Sciences, Macquarie University, Sydney, Australia. He is the author/editor of three books: *Flexible Nonparametric Curve Estimation* (2024), *Ethics in Statistics: Opportunities and Challenges* (2024), and *Practical Biostatistics for Medical and Health Sciences* (co-authored with Seyed Hassan Saneii; 2024).

Long Memory Time Series Analysis

Gnanadarsha Sanjaya Dissanayake
and Hassan Doosti

CRC CRC Press
Taylor & Francis Group
Boca Raton London New York

CRC Press is an imprint of the
Taylor & Francis Group, an **informa** business
A CHAPMAN & HALL BOOK

First edition published 2026
by CRC Press
2385 NW Executive Center Drive, Suite 320, Boca Raton FL 33431

and by CRC Press
4 Park Square, Milton Park, Abingdon, Oxon, OX14 4RN

CRC Press is an imprint of Taylor & Francis Group, LLC

ISBN: 978-1-032-62696-3 (hbk)
ISBN: 978-1-032-62699-4 (pbk)
ISBN: 978-1-032-62700-7 (ebk)

DOI: 10.1201/9781032627007

Typeset in Nimbus font
by KnowledgeWorks Global Ltd.

Contents

1

Introduction to AR, MA Time Series, Autocorrelation, Partial Autocorrelation, Spectral Density

Synopsis: This chapter provides a descriptive synthesis of time series fundamentals such as autocorrelation, partial autocorrelation, spectral density, and AR and MA processes. Exposition of knowledge in such areas begins with the basic definitions that form the cornerstones of the subject known as time series analysis. It provides the basis to learn much more advanced and complex conceptual paradigms such as long memory in the subsequent chapters.

1.1 Background

The subject area of time series analysis is used in many disciplines including finance, economics, astronomy, environmental science, medicine, physics, engineering, and hydrology. It is mainly used to infer properties of a system by the analysis of a measured time record referred to as data. This is done by fitting the best possible model to the data aiming to discover the underlying structure to an acceptable degree of accuracy. Conventional time series analysis is based on the assumptions of linearity and stationarity. However, in recent times, there has been a growing interest in nonlinear and non-stationary time series models in many practical applications. The first and the simplest reason for it is that many real-world problems do not adhere to the assumptions of linearity and/or stationarity.

In general time series analysis, it is established that there are a large number of nonlinear features such as cycles, asymmetries, bursts, jumps, chaos, thresholds, heteroscedasticity, and combinations of them that should be taken into consideration. For example, the analysis of financial markets suggests that there is a greater need to explain behaviors that are far from being even approximately linear. Therefore, the enhancement of the theory and applications for nonlinear models is essential. To consider such attributes from an analytical perspective certain basic definitions and notation on time series analysis will be useful to comprehend the material in all the chapters of this book.

DOI: 10.1201/9781032627007-1

As the analysis of time series becomes a subclass of stochastic processes, we begin with the following preliminary definitions:

1.2 Preliminaries

Definition 1.1. A stochastic process is a family of random variables $\{X_t\}$, indexed by a parameter t, where t belongs to some index set τ.

In terms of stochastic processes, the concept of *stationarity* plays an important role in many applications.

Definition 1.2. A stochastic process $\{X_t \; ; \; t \in T\}$ is said to be strictly stationary if the probability distribution of the process is invariant under translation of the index, i.e., the joint probability distributions of the index: $(X_t, \ldots, X_{t_n})$ is identical to that of $(X_{t_1+k}, \ldots, X_{t_n+k})$, for all $n \in Z^+$ (Set of positive integers), $(t_1, \ldots, t_n) \in \tau, k \in Z$ (set of integers), i.e.,

$$F(x_1, \ldots, x_n; t_1, \ldots, t_n) = F(x_1, \ldots, x_n; t_1 + k, \ldots, t_n + k). \qquad (1.1)$$

Definition 1.3. A stochastic process $\{X_t\}$ is said to be a Gaussian process if and only if the probability distribution associated with any set of time points is multivariate normal.

In particular, if the multivariate moments $E(X_{t_1}^{s_1} \ldots X_{t_n}^{s_n})$ depend only on the time differences, the process is called stationary up to order s, when $s \leq s_1 + \cdots + s_n$.

Note that, the second-order stationarity is obtained by setting $s = 2$ and this weak stationarity asserts that the mean μ is a constant (i.e., independent of t) and the covariance function $\gamma_{t\tau}$ is dependent only on the time difference. That is,

$$E(X_t) = \mu, \quad \text{for all } t$$

and

$$\text{Cov}(X_t, X_\tau) = E[(X_t - \mu)(X_\tau - \mu)] = \gamma_{|t-\tau|}, \quad \text{for all } t, \tau.$$

Time difference $k = |t - \tau|$ is called the **lag**. The corresponding **autocovariance function** is denoted by γ_k.

Definition 1.4. The acf (autocorrelation function) of a stationary process X_t is a function whose value at lag k is

$$\rho_k = \frac{\gamma_k}{\gamma_0} = \text{Corr}(X_t, X_{t+k}), \quad \text{for all } t, k \in \mathbb{Z}. \qquad (1.2)$$

Definition 1.5. The pacf (partial autocorrelation function) at lag k of a stationary process $\{X_t\}$ is the additional correlation between X_t and X_{t+k} when the linear dependence of $X_{t+1}, \ldots, X_{t+k-1}$ is removed.

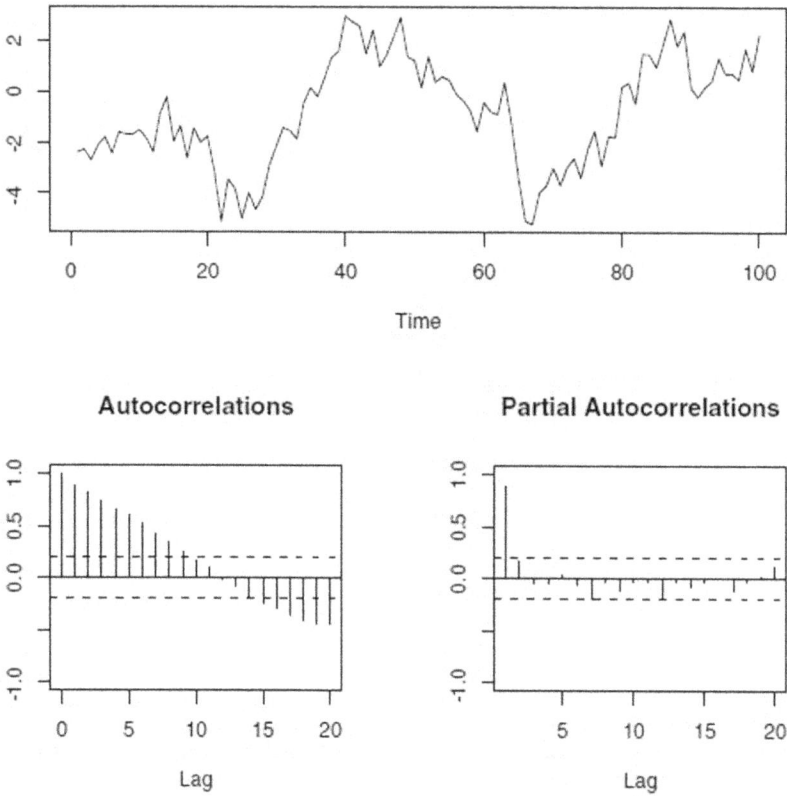

FIGURE 1.1
Time series realization, acf, and pacf of an AR(1) process.

Remark 1.1. Visual illustrations of acf and pacf are provided in Figures 1.1 and 1.2 of this chapter.

Definition 1.6. The process $\{\epsilon_t\}$ is said to be white noise (WN) with mean 0 and variance σ^2 2 if and only if ϵ_t is a sequence of uncorrelated random variables (not necessarily independent) with zero mean and autocovariance function

$$\gamma_k = \begin{cases} \sigma^2, & \text{if } k = 0, \\ 0, & \text{if } k \neq 0. \end{cases} \tag{1.3}$$

This is written as:

$$\{\epsilon_t\} \sim WN(0, \sigma^2). \tag{1.4}$$

Remark 1.2. A wide class of discrete stationary time series models can be generated by using WN as forcing terms in a set of linear difference equations.

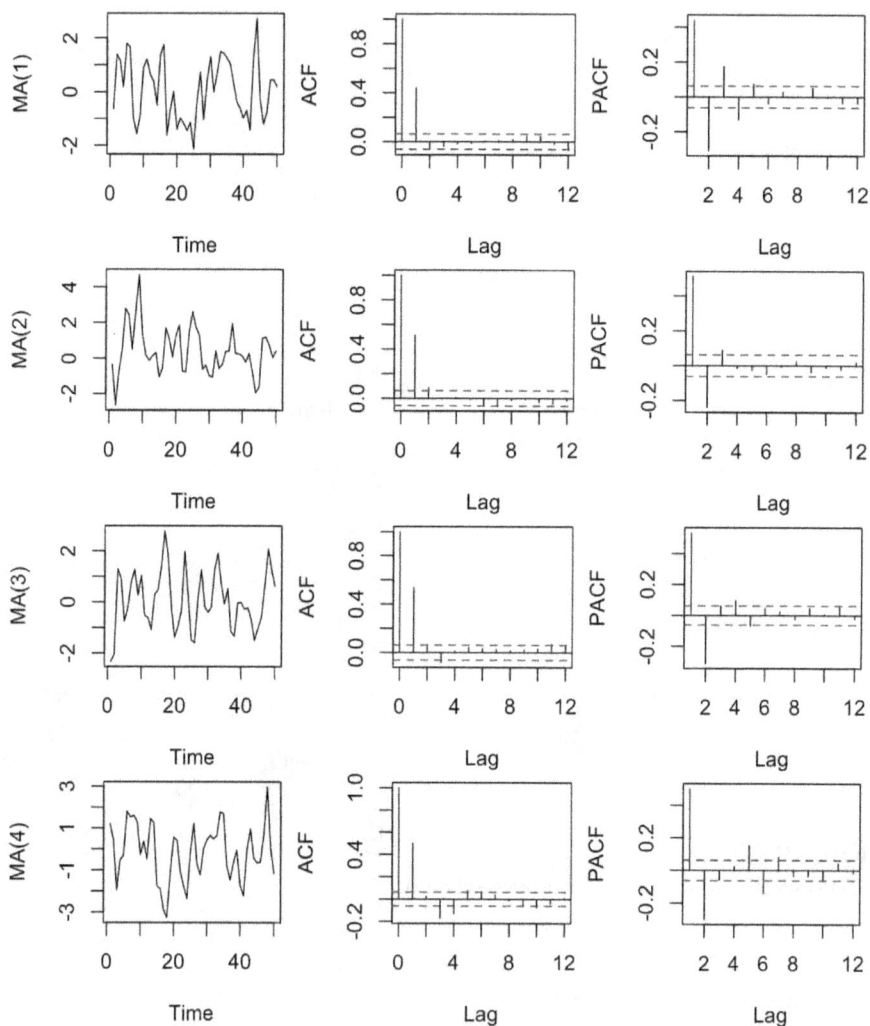

FIGURE 1.2
Time series realizations, acf, and pacf of MA processes.

If the random variables ϵ_t are independently and identically distributed (*iid*) with mean 0 and variance σ^2, then it is written

$$\{\epsilon_t\} \sim IID(0, \sigma^2). \tag{1.5}$$

Note: In this case, ϵ_t can be considered as strict WN.

Definition 1.7. Let $\{X_t\}$ be a stationary time series with autocovariance at lag k, $\gamma_k = Cov(X_t, X_{t+k}), \rho_k = Corr(X_t, X_{t+k})$, and the normalized spectrum or spectral density function (sdf) is $f(\omega) = \frac{1}{2\pi} \sum_{k=-\infty}^{\infty} \rho_k e^{-i\omega k}, \quad -\pi <$

$\omega < \pi$, where ω is the Fourier frequency. There are two main types of time series uniquely identified by the behavior of ρ_k and $f(\omega)$. They are classified as *memory types* and will be discussed in detail in Chapter 5 of this book.

Definition 1.8. A *random walk* is the process by which randomly moving objects wander away from where they started.

For example, the value of a time series at time t is the value of the series at time $t - 1$ plus a completely random movement determined by w_t. More generally, a constant drift factor δ is introduced:

$$X_t = \delta + X_{t-1} + w_t.$$

$$X_t = \delta t + \sum_{i=1}^{t} w_i.$$

Definition 1.9. A time series is a set of observations on X_t, each being recorded at a specific time t, where $t \in (0, \infty)$. In other words, a time series is a sequential set of data points, measured typically over successive times (refer: Figure 1.3 for a visual illustration and the related generating software code segment in R).

Definition 1.10. The backshift operator B is defined as:

$$BX_t = X_{t-1}.$$

It can be extended as:

$$B^2 X_t = B(BX_t) = B(X_{t-1}) = X_{t-2}.$$

Therefore,

$$BX_t = X_{t-k}$$

Similarly, the inverse (forward shift) operator could be defined by enforcing $B^{-1}B = 1$ as follows:

$$X_t = B^{-1}BX_t = B^{-1}X_{t-1}.$$

Exercise: Please run the R code given below shown in Figure 1.3 and assess the output?
Set seed for reproducibility
set.seed(123)
Create a time sequence from January 2020 to December 2022 with monthly frequency
$datesequence = seq(as.Date(\text{``}2020{-}01{-}01\text{''}), as.Date(\text{``}2022{-}12{-}31\text{''}), by = \text{``}months\text{''})$
Create a linear trend with some random noise
$lineartrend = seq(50, 150, length.out = length(datasequence))$

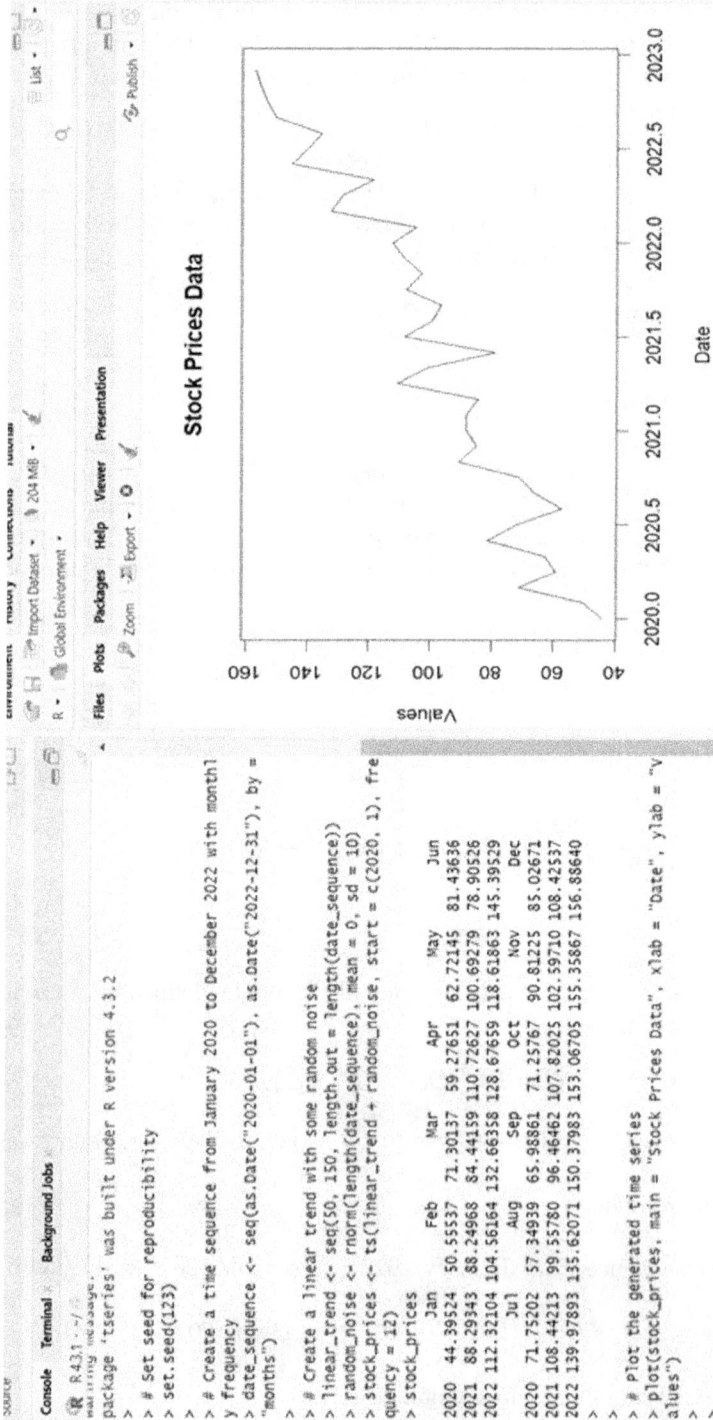

```
package 'tseries' was built under R version 4.3.2
>
> # Set seed for reproducibility
> set.seed(123)
>
> # Create a time sequence from January 2020 to December 2022 with monthly
y frequency
> date_sequence <- seq(as.Date("2020-01-01"), as.Date("2022-12-31"), by =
"months")
>
> # Create a linear trend with some random noise
> linear_trend <- seq(50, 150, length.out = length(date_sequence))
> random_noise <- rnorm(length(date_sequence), mean = 0, sd = 10)
> stock_prices <- ts(linear_trend + random_noise, start = c(2020, 1), fre
quency = 12)
> stock_prices
          Jan       Feb       Mar       Apr       May       Jun
2020  44.39524  50.55537  71.30137  59.27651  62.72145  81.43636
2021  88.29343  88.24968  84.44159 110.72627 100.69279  78.90526
2022 112.32104 104.56164 132.66358 128.67659 118.61863 145.39529
          Jul       Aug       Sep       Oct       Nov       Dec
2020  71.75202  57.34939  65.98861  71.25767  90.81225  85.02671
2021 108.44213  99.55780  96.46462 107.82025 102.59710 108.42537
2022 139.97893 135.62071 150.37983 153.06705 155.35867 156.85640
>
> # Plot the generated time series
> plot(stock_prices, main = "Stock Prices Data", xlab = "Date", ylab = "v
alues")
>
>
```

FIGURE 1.3

Time series realization plot and generating R code.

$randomnoise = rnorm(length(datesequence), mean = 0, sd = 10)$
$stockprices = ts(lineartrend+randomnoise, start = c(2020, 1), frequency =$
12)
stockprices
Plot the generated time series
$plot(stockprices, main = \text{"StockPricesData"}, xlab = \text{"Date"}, ylab =$
"Values")

1.3 Components of a Time Series

In general, a time series is affected by four components. They are trends, cycles, seasonality, and irregular (random) variation (also known as errors, residuals, noise, or innovations).

1.3.1 Trends $T(t)$

The general tendency of a time series to increase, decrease, or stagnate over a long period of time is known as a *trend*.

1.3.2 Cycles $C(t)$

This component describes medium-term changes caused by circumstances, which repeat in cycles. The duration of a cycle extends over longer period of time.

1.3.3 Seasonality $S(t)$

This component explains fluctuations within a year during the season, usually caused by climate and weather conditions, customs, traditional habits, etc.

1.3.4 Irregular variations $I(t)$

Irregular or random variations in a time series are caused by unpredictable influences, which are not regular and also do not repeat in a particular pattern.

These variations are caused by incidences such as war, strike, earthquake, flood, and revolution. There is no defined statistical technique for measuring random fluctuations in a time series.

1.3.5   Combination of the components

Considering the effects of these four components, two different types of models are generally used for a time series. They are additive and multiplicative models.

1.3.6   Additive model

$$X(t) = T(t) + C(t) + S(t) + I(t)$$

Assumption: These four components are independent of each other.

1.3.7   Multiplicative model

$$X(t) = T(t) \cdot C(t) \cdot S(t) \cdot I(t)$$

Assumption: These four components of a time series are not necessarily independent, and they can affect one another. They could be components of different categories of time series that are presented in the next subsection with relevant terminology.

1.4   Categories and Terminologies

1.4.1   Time domain versus frequency domain

Time-domain approach: How does what happened today affect what will happen tomorrow?

These approaches view the investigation of *lagged relationships* as most important, for example, autocorrelation analysis.

Frequency-domain approach: What is the economic cycle through periods of expansion and recession?

These approaches view the investigation of cycles as most important. Examples are spectral analysis and wavelet analysis.

1.4.2   Univariate versus multivariate

A time series containing records of a single variable is termed as univariate, but if records of more than one variable are considered, then it is termed as multivariate.

1.4.3   Linear versus nonlinear

A time series model is said to be linear or nonlinear depending on whether the current value of the series is a linear or nonlinear function of past observations.

1.4.4 Discrete versus continuous

In a continuous time series, observations are measured at every instance of time, whereas a discrete time series contains observations measured at discrete points in time.

The background information provided in this section forms the basis in conducting analytical time series research and its evolution. In lieu of it next discussion, compare and contrast different memory types of time series.

Time series analysis comprises methods for analyzing time series data in order to extract meaningful statistics and other characteristics of the data.

The advent of autoregressive (AR) and moving average (MA) time series were based on the conceptual paradigms presented earlier and will become the focal point of the next subsection.

1.5 AR and MA Time Series

1.5.1 AR time series

AR models are based on the idea that the current value of the series, X_t, can be explained as a linear combination of p past values, $X_{t-1}, X_{t-2}, \ldots, X_{t-p}$, together with a *random error* in the same series.

An AR model of order p, abbreviated AR(p), is of the form:

$$X_t = \phi_1 X_{t-1} + \phi_2 X_{t-2} + \cdots + \phi_p X_{t-p} + w_t = \sum_{i=1}^{p} \phi_i X_{t-i} + w_t, \qquad (1.6)$$

where X_t is stationary, $w_t \sim WN(0, \sigma_w^2)$, and $\phi_1, \phi_2, \ldots, \phi_p$ (with $\phi_p \neq 0$) are model parameters. The hyperparameter p represents the length of the series.

By using the backshift operator, we can rewrite equation (1.6) as:

$$X_t - \phi_1 X_{t-1} - \phi_2 X_{t-2} - \cdots - \phi_p X_{t-p} = w_t = (1 - \phi_1 B - \phi_2 B^2 - \cdots - \phi_p B^p) X_t, \qquad (1.7)$$

where the AR operator is defined as:

$$\phi(B) = 1 - \phi_1 B - \phi_2 B^2 - \cdots - \phi_p B^p = 1 - \sum_{j=1}^{p} \phi_j B^j.$$

Therefore, AR(p) can be concisely written as:

$$\phi(B) X_t = w_t. \qquad (1.8)$$

The simplest AR process is AR(0), which has no dependence between the terms. In fact, AR(0) is essentially *white noise*.

AR(1) process is generally represented as $X_t = \phi_1 X_{t-1} + w_t$. Only the previous term in the process and the noise term contribute to the output.

If $|\phi_1|$ is close to zero, the process still looks like WN.

If $\phi_1 < 0$, X_t tends to oscillate between positive and negative values.

If $\phi_1 = 1$, the process becomes equivalent to a random walk.

Example: For AR(1) processes: $X_t = 0.9X_{t-1} + w_t$ and $X_t = -0.9X_{t-1} + w_t$, the mean $= E[X_t] = 0$, variance $Var(X_t) = \frac{\sigma_w^2}{1-\phi_1^2}$.

1.5.2   AR models – parameter estimation

There are many feasible approaches to parameter estimation of AR models. Some of the popular methods are:

1. Method of moments estimation (e.g., Yule-Walker estimator),

2. Maximum-likelihood estimation (MLE) estimator,

3. Ordinary least square (OLS) estimator.

1.5.3   MA time series

Name might be misleading, but MA time series models should not be confused with MA smoothing.

Recall that in AR models, the current observation X_t is regressed using the previous observations $X_{t-1}, X_{t-2}, \ldots, X_{t-p}$, plus an error term w_t at the current time point.

One problem with the AR model is the ignorance of correlated noise structures (which are unobservable) in the time series. In other words, the imperfectly predictable terms in current time, w_t, and previous steps, $w_{t-1}, w_{t-2}, \ldots, w_{t-q}$, are also *informative* for predicting observations.

MA model of order q, or MA(q), is defined as:

$$X_t = w_t + \theta_1 w_{t-1} + \theta_2 w_{t-2} + \cdots + \theta_q w_{t-q} = w_t + \sum_{j=1}^{q} \theta_j w_{t-j}, \qquad (1.9)$$

where $w_t \sim WN(0, \sigma_w^2)$ and $\theta_1, \theta_2, \ldots, \theta_q$ (with $\theta_q \neq 0$) are model parameters. The hyper parameter q represents the length of the series. Figure 1.2 provides visual illustrations of series realization, acf, and pacf for MA series having four different values of q.

Although it looks like a regression model, the difference is that w_t is not observable. Contrary to the AR model, a finite MA model is always stationary because the observation is just a weighted MA over past forecast errors.

Therefore, the MA operator is defined as:

$$\theta(B) = 1 + \theta_1 B + \theta_2 B^2 + \cdots + \theta_q B^q,$$

where B stands for the backshift operator, and $B(w_t) = w_{t-1}$.

Therefore, the MA model can be re-written as:

$$X_t = w_t + \theta_1 w_{t-1} + \theta_2 w_{t-2} + \cdots + \theta_q w_{t-q} = (1 + \theta_1 B + \theta_2 B^2 + \cdots + \theta_q B^q) w_t.$$
(1.10)

Therefore, MA(q) can be concisely written as:

$$X_t = \theta(B) w_t. \qquad (1.11)$$

Examples: For a simulated MA(1) process – $X_t = w_t + 0.8 w_{t-1}$, mean $E[X_t] = 0$, variance $\text{Var}(X_t) = \sigma_w^2(1 + \theta_1^2)$.

For a simulated MA(2) process – $X_t = w_t + 0.5 w_{t-1} + 0.3 w_{t-2}$, mean $E[X_t] = 0$, variance $\text{Var}(X_t) = \sigma_w^2(1 + \theta_1^2 + \theta_2^2)$.

In fact, all causal AR(p) processes can be represented as MA(∞); in other words, infinite MA processes are finite AR processes.

All invertible MA(q) processes can be represented as AR(∞); i.e., finite MA processes are infinite AR processes.

Combining of AR and MA processes results in the generation and creation of revolutionary ARMA time series, which becomes the topic of discussion in the next chapter.

Note: Some of the material presented in this chapter could be found in Dissanayake (2015).

1.6 Chapter 1 Questions

1. Does the sample PACF of a time series always corroborate its ACF as a visual diagnostic tool?

2. What are the four main components linked with time series?

3. Provide the generic equations of the two types of time series models that could be generated by combining the four main components.

4. What are the various categories linked with time series?

5. What are the types of errors (random or irregular variations or residuals or innovations) introduced in the chapter?

6. Conventional time series analysis is based on what assumptions?

7. Name three types of methodologies that could be used for AR parameter estimation.

8. List the differences between AR and MA time series processes.

9. What is a random walk?

10. Describe the link between covariance, correlation, and spectral density at a predefined lag k.

2

ARMA Process and Box–Jenkins Model

Synopsis: This chapter provides an extension of the material presented in the preceding chapter. The AR and MA time series introduced in Chapter 1 will be linked together to form the autoregressive–moving-average (ARMA) process time series and the related fundamentals of the introduced model will be discussed in detail. Thereafter, an important extension of the ARMA process known as the Box–Jenkins model would become the topic of interest toward the latter part of the chapter.

2.1   Background

In time series analysis, ARMA models provide a parsimonious description of a (weakly) stationary stochastic process in terms of two polynomials, one for the autoregression (AR) and the second for the moving average (MA). The general ARMA model was described in the 1951 thesis of Peter Whittle, hypothesis testing in time series analysis, and it was popularized in the 1970 book by George E. P. Box and Gwilym Jenkins. An ARMA model is a tool for understanding and, perhaps, predicting future values in this series. The AR part involves regressing the variable on its own lagged (i.e., past) values. The MA part involves modeling the error term as a linear combination of error terms occurring contemporaneously and at various times in the past. The model is usually referred to as the ARMA(p, q) model, where p is the order of the AR part and q is the order of the MA part (as defined below). ARMA models can be estimated by using the Box–Jenkins method.

2.2   Preliminaries

The notation ARMA (p, q) refers to the model with p autoregressive terms and q moving-average terms. This model contains the AR(p) and MA(q) models,

DOI: 10.1201/9781032627007-2

such that

$$X_t = \epsilon_t + \sum_{i=1}^{p} \phi_i X_{t-i} + \sum_{i=1}^{p} \theta_i \epsilon_{t-i}, \qquad (2.1)$$

The general ARMA model was described in the 1951 thesis of Peter Whittle, who used mathematical analysis (Laurent series and Fourier analysis) and statistical inference. ARMA models were popularized by a 1970 book by George E. P. Box and Jenkins, who expounded an iterative (Box–Jenkins) method for choosing and estimating them. This method was useful for low-order polynomials (of degree three or less).

The ARMA model is essentially an infinite impulse response filter applied to white noise, with some additional interpretation placed on it.

ARMA model could be interpreted using the lag operator B employing the following deriving steps:

AR(p) model is given as:

$$\epsilon_t = (1 - \sum_{i=1}^{p} \phi_i B^i) X_t) = \phi(B) X_t \qquad (2.2)$$

where $\phi(B) = 1 - \sum_{i=1}^{p} \phi_i B^i$.

Similarly, the MA(q) model is given as:

$$X_t - \mu = (1 + \sum_{i=1}^{q} \theta_i B^i) \epsilon_t) = \theta(B) \epsilon_t \qquad (2.3)$$

where $\theta(B) = 1 + \sum_{i=1}^{q} \theta_i B^i$.

Therefore, the final ARMA model could be given as:

$$(1 - \sum_{i=1}^{p} \phi_i B^i) X_t) = (1 + \sum_{i=1}^{q} \theta_i B^i) \epsilon_t \qquad (2.4)$$

More concisely, it could be given as: $\phi(B) X_t = \theta(B) \epsilon_t$ or as $\frac{\phi(B)}{\theta(B)} X_t = \epsilon_t$.

Another alternative notation for the ARMA process begins with the polynomials involving the lag operator to appear in a similar form continuously as follows:

$$(1 - \sum_{i=1}^{p} \phi_i B^i) X_t = (1 + \sum_{i=1}^{q} \theta_i B^i) \epsilon_t$$

Beginning the summations with $i = 0$, $\phi_0 = -1$, and $\theta_0 = 1$ results in the creation of the following elegant equation:

$$-\sum_{i=0}^{p} \phi_i B^i = \sum_{i=1}^{q} \theta_i B^i \epsilon_t$$

Note: In digital signal processing, ARMA model becomes a digital filter with white noise as the input and the ARMA process as the output result.

2.2.1   Fitting an ARMA model

2.2.1.1   Choosing model parameters p and q

Selecting suitable values of p and q in the $\text{ARMA}(p, q)$ model can be achieved by plotting the partial autocorrelation function for an estimate of p, and likewise using the autocorrelation function for an estimate of q. A visual illustration of the two functions is provided in Figure 2.1. Extended autocorrelation functions (EACF) can be used to simultaneously determine p and q. Further information can be comprehended by considering the same functions for the residuals of a model fitted with an initial selection pair of estimates for p and q.

Brockwell and Davis (1991) recommends using the Akaike information criterion (AIC) for finding p and q. Another alternative choice for order determining is the BIC criterion.

2.2.1.2   Estimating model parameter coefficients of p and q

After the selection of p and q in general ARMA models can be fitted using the least squares regression method to find values of the parameters that will minimize the error term. Generally it is considered a good practice to find the smallest values of p and q that will provide an acceptable fit to the data. On the contrary, for a pure AR model the Yule–Walker equations may be used to obtain a fit.

Unlike other regression methods, such as OLS and 2SLS, often employed in statistical and econometric analysis, ARMA model outputs are used mainly for forecasting time-series data. As such, their coefficients are only utilized for prediction. The coefficients should then be seen only as useful for predictive modeling.

2.2.1.3   Implementation of ARMA process using statistical packages

In R, the arima function (in standard package stats) is documented in ARIMA Modeling of Time Series. The package astsa has an improved script called sarima for fitting ARMA models (seasonal and nonseasonal) as well as sarima.sim to simulate data from these models. Extension packages contain related and extended functionality, e.g., the tseries package includes an arma function, documented in "Fit ARMA Models to Time Series"; the fracdiff package contains fracdiff() for fractionally integrated ARMA processes; and the forecast package includes auto.arima for selecting a parsimonious set of p, q. The CRAN task view on Time Series contains links to most of these.

MATLAB includes functions such as arma, ar, and arx to estimate AR, ARX (autoregressive exogenous), and ARMAX models. See System Identification Toolbox and Econometrics Toolbox for more information.

PyFlux has a Python-based implementation of ARIMAX models, including Bayesian ARIMAX models.

ARMA: Autocorrelation (left) and Partial Autocorrelation (right)

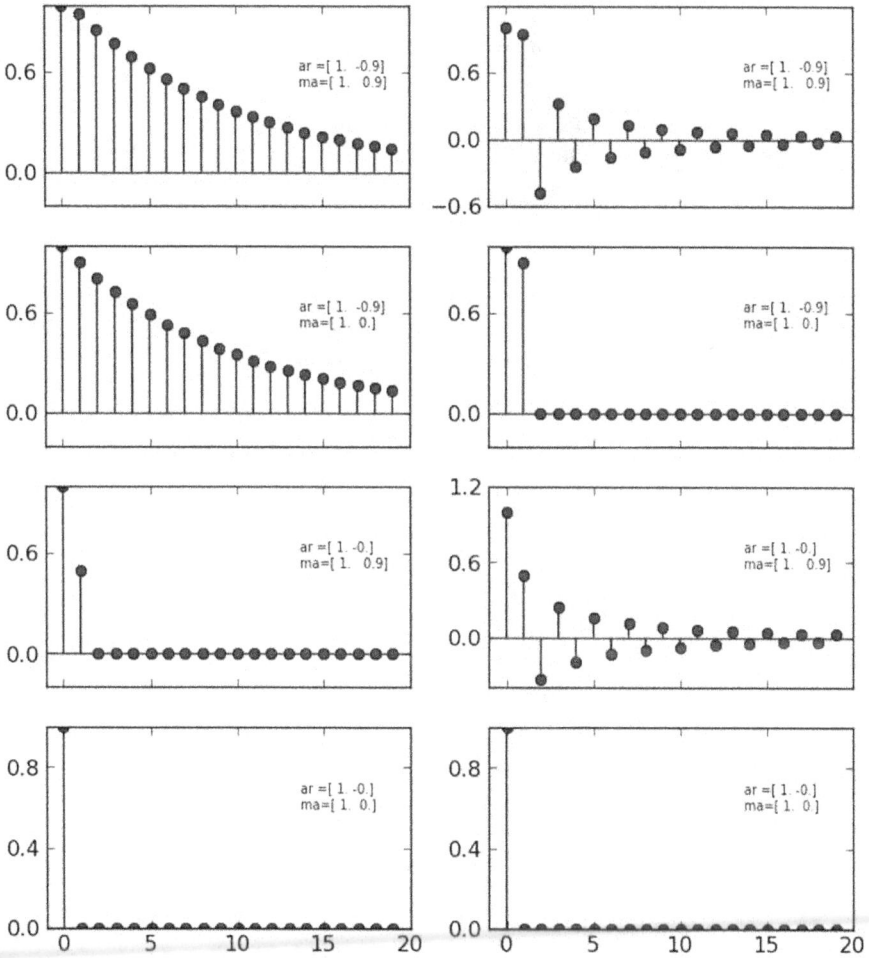

FIGURE 2.1
Images of ARMA ACF and PACF.

Stata includes the function arima which can estimate ARMA and ARIMA models.

SAS has an econometric package, ETS, that estimates ARIMA models.

2.2.1.4   Spectral density of an ARMA process

The spectral density of an ARMA process is

$$S(\omega) = \frac{\sigma^2}{2\pi} \{ \frac{\mid \theta(e^{-i\omega}) \mid}{\mid \phi(e^{-i\omega}) \mid} \}^2 \tag{2.5}$$

where σ^2 is the variance of white noise, ϕ is the characteristic polynomial of the AR process, and θ is the characteristic polynomial of the MA process.

2.2.1.5   Applications of an ARMA process

An ARMA process is appropriate when a system is a function of a series of unobserved shocks (MA or moving average part) as well as its own behavior. As a case in point, stock prices may be shocked by fundamental information as well as exhibiting technical trending and mean-reversion effects due to market participants.

Note: ARMA model is a univariate model. Extensions for the multivariate case are the vector autoregression (VAR) and vector autoregression moving-average (VARMA) models.

2.2.1.6   Autoregressive–moving-average process with exogenous inputs model (ARMAX model)

Notation ARMAX(p, q, b) refers to a model with p autoregressive terms, q moving average terms and b exogenous inputs terms. This model contains the AR(p) and MA(q) models and a linear combination of the last b terms of a known and external time series d_t.

It is given by

$$X_t = \epsilon_t + \sum_{i=1}^{p} \phi_i X_{t-i} + \sum_{i=1}^{q} \theta_i \epsilon_{t-i} + \sum_{i=1}^{b} \eta_i d_{t-i} \tag{2.6}$$

where $\eta_1, ..., \eta_b$ are parameters of exogeneous input d_t.

Statistical packages do implement the ARMAX model through the usage of exogeneous (independent) variables. Need to be extremely cautious, when interpreting the output of those packages, because the estimated parameters usually (for example, in R) refer to regression given as:

$$X_t - m_t = \epsilon_t + \sum_{i=1}^{p} \phi_i (X_{t-i} - m_{t-i}) + \sum_{i=1}^{q} \theta_i \epsilon_{t-i} \tag{2.7}$$

where m_t incorporates all exogeneous (independent) variables such that

$$m_t = c + \sum_{i=0}^{b} \eta_i d_{t-i}.$$

Note: One of the most important developments in time series analysis that links ARMA processes with its extended ARIMA configuration (main topic of focus in the next chapter) is the Box–Jenkins model and will become the focal point of the final segment of this current chapter.

2.2.1.7 Box–Jenkins model

In statistics and econometrics, Box–Jenkins method, named after the statisticians George Box and Gwilym Jenkins, applies the ARMA process or its related autoregressive integrated moving average (ARIMA) model to find the best fit of a time-series model to past values of a time series.

2.2.1.8 Modeling of Box–Jenkins model – the approach

Original Box–Jenkins process utilizes an iterative three-stage modeling approach:

2.2.1.9 Model identification and model selection

Making sure variables are stationary, identifying seasonality in the dependent series (seasonally differencing if necessary), using plots autocorrelation (ACF) and partial autocorrelation (PACF) functions of dependent time series to decide which (if any) autoregressive or moving average component should be used in the model.

Parameter estimation through computation algorithms to arrive at coefficients that best fit the chosen ARMA/ARIMA model. The most common methods used are maximum likelihood estimation and nonlinear least-squares estimation.

Statistical model checking by testing if the estimated model conforms to specifications of a stationary univariate process. In particular, the residuals should be independent and constant in mean and variance over time. (Plotting the mean and variance of residuals over time, performing a Ljung–Box test or plotting autocorrelation and partial autocorrelation of the residuals are required to identify misspecification.) If estimation is inadequate, we revert back to step one and attempt to build a better model.

The data used initially to run the Box–Jenkins model were from a gas furnace. These data are well known as the Box and Jenkins gas furnace data for benchmarking predictive models.

Commandeur and Koopman (2007) argue that the Box–Jenkins approach is fundamentally tedious. Problem arises because in "the economics and social fields, real series are never stationary however much differencing is done". Thus, the investigator has to address the question: how close to stationary is close enough? As the authors note, "It is a hard question to answer". Authors further argue that rather than Box–Jenkins, it is better to use state space methods, as stationarity of time series is then not required.

2.2.2   Identification Box–Jenkins model

2.2.2.1   Stationarity and seasonality

The initial step in developing a Box–Jenkins model is to determine whether the time series is stationary and if there is any significant seasonality that needs to be modeled?.

2.2.2.2   Detecting stationarity

Stationarity can be assessed from a run sequence plot. The run sequence plot should show constant location and scale. It can also be detected from an autocorrelation plot. Specifically, nonstationarity is often indicated by an autocorrelation plot with very slow decay. One can also utilize a Dickey–Fuller test or augmented Dickey–Fuller test.

2.2.2.3   Detecting seasonality

Seasonality (or periodicity) can be assessed from an autocorrelation plot, a seasonal subseries plot, or a spectral density plot.

2.2.3   Differencing to achieve stationarity

Box and Jenkins recommend differencing approach to achieve stationarity. However, fitting a curve and subtracting fitted values from original data can also be used in the context of Box–Jenkins models.

2.2.3.1   Seasonal differencing

At the model identification stage, goal is to detect existing seasonality and to identify the order for seasonal autoregressive and seasonal moving average terms. In many series, period is known and a single seasonality term is sufficient. For example, for monthly data one would typically include either a seasonal AR 12 term or seasonal MA 12 term. In using Box–Jenkins models, one does not explicitly remove seasonality before fitting the model. Instead, one includes order of the seasonal terms in the model specification to the ARIMA estimation software. However, it may be helpful to apply a seasonal difference to the data and regenerate the autocorrelation and partial auto-correlation plots. This may help in model identification of the nonseasonal component. In certain cases, seasonal differencing may remove most or all of the seasonality effect.

2.2.3.2   Identify p and q

Once stationarity and seasonality have been addressed, next step will be to identify the order (p and q) of the autoregressive and moving average terms. Different authors have different approaches in identifying p and q. Brock-well and Davis (1991) state "our prime criterion for model selection [among

ARMA(p, q) models] will be the AICc", i.e., the Akaike information criterion with correction. Other authors use the autocorrelation plot and the partial autocorrelation plot, described below.

2.2.3.3   Autocorrelation and partial autocorrelation plots

Sample autocorrelation and sample partial autocorrelation plots are compared to the theoretical behavior of these plots when the order is known.

Specifically, for an AR(1) process, sample autocorrelation function should have an exponentially decreasing appearance. However, higher-order AR processes are often a mixture of exponentially decreasing and damped sinusoidal components.

For higher-order autoregressive processes, sample autocorrelation needs to be supplemented with a partial autocorrelation plot. The partial autocorrelation of an AR(p) process becomes zero at lag $p+1$ and greater, so we examine the sample partial autocorrelation function to see whether there is evidence of a departure from zero. This is usually determined by placing a 95 percent confidence interval on the sample partial autocorrelation plot (most software programs that generate sample autocorrelation plots also plot the confidence interval). If the software program does not generate the confidence band, it is approximately $\pm 2/\sqrt{N}$, with N denoting the sample size.

The autocorrelation function of a MA(q) process becomes zero at lag $q+1$ and greater, so we examine the sample autocorrelation function to see where it essentially becomes zero. We do this by placing the 95 percent confidence interval for the sample autocorrelation function on the sample autocorrelation plot. Most software that can generate the autocorrelation plot can also generate this confidence interval.

The sample partial autocorrelation function is generally not helpful for identifying the order of the moving average process.

2.2.4   Box–Jenkins model estimation

Estimating parameters of Box–Jenkins models involve numerically approximating the solutions of nonlinear equations. For this reason, it is common to use statistical software packages tailor made to handle to the approach– virtually all modern statistical packages feature this capability. Main approaches to fitting Box–Jenkins models are nonlinear least squares and maximum likelihood estimation. Maximum likelihood estimation is generally the sought after technique. The likelihood equations for the full Box–Jenkins model are complex and are not included in this chapter. See Brockwell and Davis (1991) for mathematical and statistical details.

2.2.5   Box–Jenkins model diagnostics

2.2.5.1   Assumptions for a stable univariate process

Model diagnostics for Box–Jenkins models are similar to model validation measures for nonlinear least squares fitting.

That is, the error term A_t is assumed to follow assumptions of a stationary univariate process. Residuals should be white noise (or independent when their distributions are normal) drawings from a fixed distribution with a constant mean and variance. If Box–Jenkins model is a good model for the data, the residuals should satisfy the given assumptions.

If these assumptions are not satisfied, one needs to fit a much more appropriate model. That is, go back to the model identification step and try to develop a much better model. Hopefully, the analysis of residuals could provide some clues towards a more appropriate model.

One way to assess whether the residuals from the Box–Jenkins model follow the assumptions would be to generate statistical graphics (including an autocorrelation plot) of the residuals. One could also look at the Ljung–Box statistic.

Development of the Box–Jenkins model paved the way toward statisticians and econometricians conducting deeper analytical probing mechanisms with respect to integer differencing and the autoregressive integrated moving average (ARIMA) model, which becomes the main topic of interest in the next chapter.

2.3   Chapter 2 Questions

1. What are the characteristic polynomials of an ARMA process? Show the mathematical/statistical illustrations.

2. What would be the role of an ARMA model in digital signal processing?

3. What kind of decay is visible in an ACF plot of an ARMA process?

4. What kind of decay is visible in an PACF plot of an ARMA process?

5. State 2 methods of model fit that could be used to find the orders of ARMA parameters p and q?

6. Write a code in R to generate an ARMA(1,1) process and its associated ACF?

7. Write a code in R to create an ARMA(2,1) process and its associated PACF?

8. Write a code in R to generate a visual plot of the spectral density linked with the ARMA process in question 6 above?

9. Write a code in R to generate a visual plot of the spectral density linked with the ARMA process in question 6 above?

10. Create a stationary Box–Jenkins model of your choice and assess serial correlation?

3

Integer Differencing and ARIMA Process with White Noise

Synopsis: This chapter provides an extension of the material presented in the previous two chapters. Integer differencing is introduced as a concept, and it is incorporated into the ARMA model extending it to an ARIMA process. In simpler terms, an integer differencing filter is embedded into an ARMA process to create an ARIMA time series. Fundamentals in terms of theory linked with the notions of integer differencing and an ARIMA process will be discussed in detail within this chapter. Thereafter, empirical evidence corroborating the theory will become the focus in latter chapter sections.

3.1 Background

Integer differencing was introduced into time series literature to simplify complex as well as cumbersome time series models. It paved the way to transform nonstationary time series into workable, consistent processes such as the ARIMA process. It was a revolutionary development in time series-based research and in the current context has formed the basis to formulate other important, deep rooted analytical concepts and paradigms related to time series.

ARIMA models can be estimated by using integer differencing as an initial step.

3.2 Integer Differencing

Many predictive models need a certain level of consistency in terms of time series called stationarity. Usual transformation is done through integer order differencing or better known as integer differencing (in Finance, e.g., modeling returns instead of absolute prices), eliminates memory in the data and thus affects the predictive power of modeling.

DOI: 10.1201/9781032627007-3

In a generic sense, a given time series is classified as a collection of data points with respect to time generated by a stochastic process whose distribution and statistics infer a predictive model.

Creating predictive models of stochastic processes imply finding a balance between specificity and generalizability of samples: model will interpret a given series against the backdrop of generic patterns.

In terms of being specific, a time series comes with inherent ordering due to its temporal structure more than a generic predictive regression model — at any given instance it reflects a history of values traversed in the past, as well as the specific memory of its past track record.

3.2.1 Stationarity

To identify a generalized pattern of the generating process and map the given configuration, this specific series memory is most of the time eliminated as a step in preprocessing prior to actual modeling.

In terms of supervised learning in machine learning, it does serve to discover the generalized structure and match it to a majority of samples within a labeled training set.

In terms of the statistical properties of the process, the series ensemble in the form of mean, variance and covariance, should be invariant with respect to time ordering, meaning the series should not have a trend over time. Such a concept is defined as stationarity.

There are various ways to assess a time series for stationarity. They are as follows:

1. Visually inspect line plots over time for a defined trend.

2. Compare basic descriptive summary statistics (such as mean, variance, and covariance) and check for various (random) splits within the series.

3. Probe the autocorrelation plot: faster the curve drops for increasing lags less the order of nonstationarity.

4. Most common statistical test for stationarity is the augmented Dickey–Fuller (ADF) test for unit roots.

Therefore, the implication of a unit root is that formally a solution of the process characteristic equation lying on the unit circle implies that initial conditions or external shocks do not disapear over time, but extend through the series and provide successive values.

At a given confidence level, ADF tests the null-hypothesis in terms of the existence of a unit root of some order' (showing nonstationarity of the time series) against an alternative of stationarity (or in a strictly sense trend stationarity).

It will be a rather straight-forward technique to prove that the existence of a unit root does imply nonstationarity of the series.

For example: assume that $x_t = x_{t-1} + \epsilon_t$, then $x_t = x_0 + \sum_{j=0}^{t} \epsilon_j$ and $Var(x_t) = t.Var(\epsilon_j)$=t.constant.

Therefore, the variance becomes time-dependent.

Notably, a necessary assumption for many classical model type approaches will be the stationarity of a time series: i.e., if a clear trend or seasonality is present in the data, it would be necessary to remove them and model the resultant series. Thereafter, for a forecast, it would require to combine the (deterministic) trend and the model output.

Most common transformation to make a series stationary is differencing up to some order: first-order differencing is simply subtracting from each value the preceding one. To difference data, the magnitude of change between consecutive observations is computed. Mathematically, this is shown as:

$$x_t^{'} = x_t - x_{t-1}. \tag{3.1}$$

Second-order differencing repeats the process for a resulting series and for higher orders. Mathematically, it could be shown as:

$$x_t^* = x_t^{'} - x_{t-1}^{'} = (x_t - x_{t-1}) - (x_{t-1} - x_{t-2}) = x_t - 2x_{t-1} + x_{t-2}. \tag{3.2}$$

Yet, another method of differencing data is seasonal differencing. It involves computing the difference between an observation and the corresponding observation in the preceding season, e.g., a year. It is shown as:

$$x_t^{'} = x_t - x_{t-m}, \tag{3.3}$$

where m is seasonal duration.

For example, in financial time series, you would consider (log) returns instead of absolute prices to make the model fall in line to a specific price level (in most cases, first-order differencing would be adequate to ensure stationarity).

3.3   ARIMA Process

In subjects such as statistics and econometrics, and especially in time series analysis, an ARIMA model is a generalization of the ARMA model. ARIMA models are applied in certain instances where data depict nonstationarity in the sense of expected value (but not variance/autocovariance), where a first differencing step (linked with the "integrated" part of the model) can be applied one or more times to eliminate nonstationarity of the mean function (in other words the trend). When seasonality appears in a time series, seasonal differencing could be applied to eliminate the seasonal component. Since the ARMA model, according to the Wold's decomposition theorem (which will be introduced in a subsequent chapter of this text) is theoretically sufficient to describe a regular (or purely nondeterministic) wide-sense stationary time series, a motivation would arise to make stationary a nonstationary time series, for example by using differencing, before using the ARMA model. If the time series contains a predictable sub-process (known as a pure sine or complex-valued exponential process), then the predictable component is treated as a non-zero-mean yet periodic (seasonal) component in the ARIMA framework so that it will get eliminated through seasonal differencing.

AR part of ARIMA denotes that the evolving variable of interest is regressed on its own lagged (prior) values. MA part depicts that the regression error is actually a linear combination of error terms whose values did occur contemporaneously at various times in the past. The I (for "integrated") in the acronym ARIMA indicates data values have been replaced with the difference between their values and the previous values (with the differencing process performed more than once on certain occasions). Objective of such features is to make the model fit the data in an optimal manner.

Non seasonal ARIMA models are denoted ARIMA(p, d, q), where parameters p, d, and q are non-negative integers, p is the order (number of time lags) of the AR model, d is the degree of differencing (the number of times the data have had past values subtracted), and q is the order of the MA model. Seasonal ARIMA models (that will be discussed in detail in a subsequent chapter of the text) are usually denoted ARIMA$(p, d, q)(P, D, Q)m$, where m refers to the number of periods in each season, and the uppercases P, D, Q refer to the autoregressive, differencing, and moving average terms for the seasonal part of the ARIMA model.

ARIMA models can be estimated following the Box–Jenkins approach.

An ARIMA(p, d, q) process expressed in terms of the polynomial factorization property by

$$(1 - \sum_{i=1}^{p} \phi_i L^i)(1 - L)^d X_t = (1 + \sum_{i=1}^{q} \theta_i L^i)\epsilon_t,$$

where L is the lag operator and $(1 - L)^d$ becomes the integrating filter with a differencing parameter d as an integer and thus can be thought of as a particular case of an ARMA$(p + d, q)$ process having an AR polynomial with d unit roots. (For this reason, no process that is accurately described by an ARIMA model with $d > 0$ is wide-sense stationary.)

The equation given above can be generalized as follows:

$$(1 - \sum_{i=1}^{p} \phi_i L^i)(1 - L)^d X_t = \delta + (1 + \sum_{i=1}^{q} \theta_i L^i)\epsilon_t,$$

It is defined as an ARIMA(p, d, q) process with drift $\frac{\delta}{\sum_{i=1}^{p} \phi_i}$.

3.3.1 Other special forms

Explicit identification of the factorization of the AR polynomial into factors as above can be extended to other cases, firstly to apply to the MA polynomial and secondly to include other special factors. For example, having a factor $(1 - L^s)$ in the model is one way of including a nonstationary seasonality of period s within the model; this factor has the effect of re-expressing the data as changes from s periods ago. Another illustration is the factor $\left(1 - \sqrt{3}L + L^2\right)$, which includes a non-stationary seasonality of period 2. Effect of the first type of factor is to allow each season's value to drift separately over time, whereas with the second type values for adjacent seasons move together.

Identification and specification of appropriate factors in an ARIMA model can be an important step in modeling as it can allow a reduction in the overall number of parameters to be estimated while allowing the imposition on the model of types of behavior that logic and experience suggest should be there.

3.3.2 Examples of special ARIMA models

Certain special cases of ARIMA models arise naturally and are mathematically equivalent to other popular forecasting models. Examples are as follows:

1. ARIMA(0, 1, 0) model (or I(1) model) is given by $X_t = X_{t-1} + \epsilon_t$, which is simply a **random walk**.

2. ARIMA(0, 1, 0) with a constant, given by $X_t = c + X_{t-1} + \epsilon_t$, which is simply a **random walk with drift**.

3. ARIMA(0, 0, 0) model is a **white noise** model.

4. ARIMA(0, 1, 2) model is a **Damped Holt's** model.

5. ARIMA(0, 1, 1) model without constant is a **basic exponential smoothing** model.

6. ARIMA(0, 2, 2) model is given by $X_t = 2X_{t-1} - X_{t-2} + (\alpha + \beta - 2)\epsilon_{t-1} + (1 - \alpha)\epsilon_{t-2} + \epsilon_t$, which is equivalent to Holt's linear method with additive errors, or double exponential smoothing.

3.3.3 Choosing the order

The orders p and q can be determined using the sample autocorrelation function (ACF), partial autocorrelation function (PACF), and/or extended auto-correlation function (EACF) method.

Other alternative methods include AIC, BIC, etc. To determine the order of a nonseasonal ARIMA model, a useful criterion is the Akaike information criterion (AIC). It is written as:

$$AIC = -2log(L) + 2(p + q + k). \tag{3.4}$$

where L is the likelihood of the data, p is the order of the AR part, and q is the order of the MA part. k represents the intercept of the ARIMA model. For AIC, if $k = 1$, then there is an intercept in the ARIMA model ($c \neq 0$), and if $k = 0$, then there is no intercept in the ARIMA model ($c = 0$).

Correct AIC for ARIMA models can be written as:

$$AIC_c = AIC + \frac{2(p+q+k)(p+q+k+1)}{T-p-q-k-1}. \tag{3.5}$$

Bayesian information criterion (BIC) can be written as:

$$BIC = AIC + ((logT) - 2)(p+q+k) \tag{3.6}$$

Objective is to minimize the AIC, AICc, or BIC values for a good model. Lower the value of one of these criteria for a range of models being investigated, the better the model will suit the data. AIC and BIC are used for two completely different purposes. While the AIC tries to approximate models toward the reality of the situation, the BIC attempts to find the perfect fit. BIC approach is often criticized as there never is a perfect fit to real-life complex data; however, it is still a useful technique for selection as it penalizes models heavily for having more parameters than AIC.

AICc can be used to compare ARIMA models with the same orders of differencing only. For ARIMAs with different orders of differencing, root mean square error (RMSE) can be used for model comparison.

3.3.4 Forecasting using ARIMA models

An ARIMA model can be viewed as a "cascade" of two models. The first is nonstationary:

$$Y_t = (1 - L)^d X_t \tag{3.7}$$

While the second is wide-sense stationary:

$$\left(1 - \sum_{i=1}^{p} \phi_i L^i\right) Y_t = \left(1 + \sum_{i=1}^{q} \theta_i L^i\right) \epsilon_t,$$

Now, forecasts can be made for the process Y_t using a generalization of the method of autoregressive forecasting.

3.3.5 Extension of ARIMA model: SARIMA time series

SARIMA stands for seasonal autoregressive integrated moving average. It is a versatile and widely used robust time series forecasting model. Considered as an extension of the nonseasonal ARIMA model to handle data with seasonal patterns. A special feature of the SARIMA process is that it captures both the short- and long-term dependencies within the data when utilized to generate forecast projections.

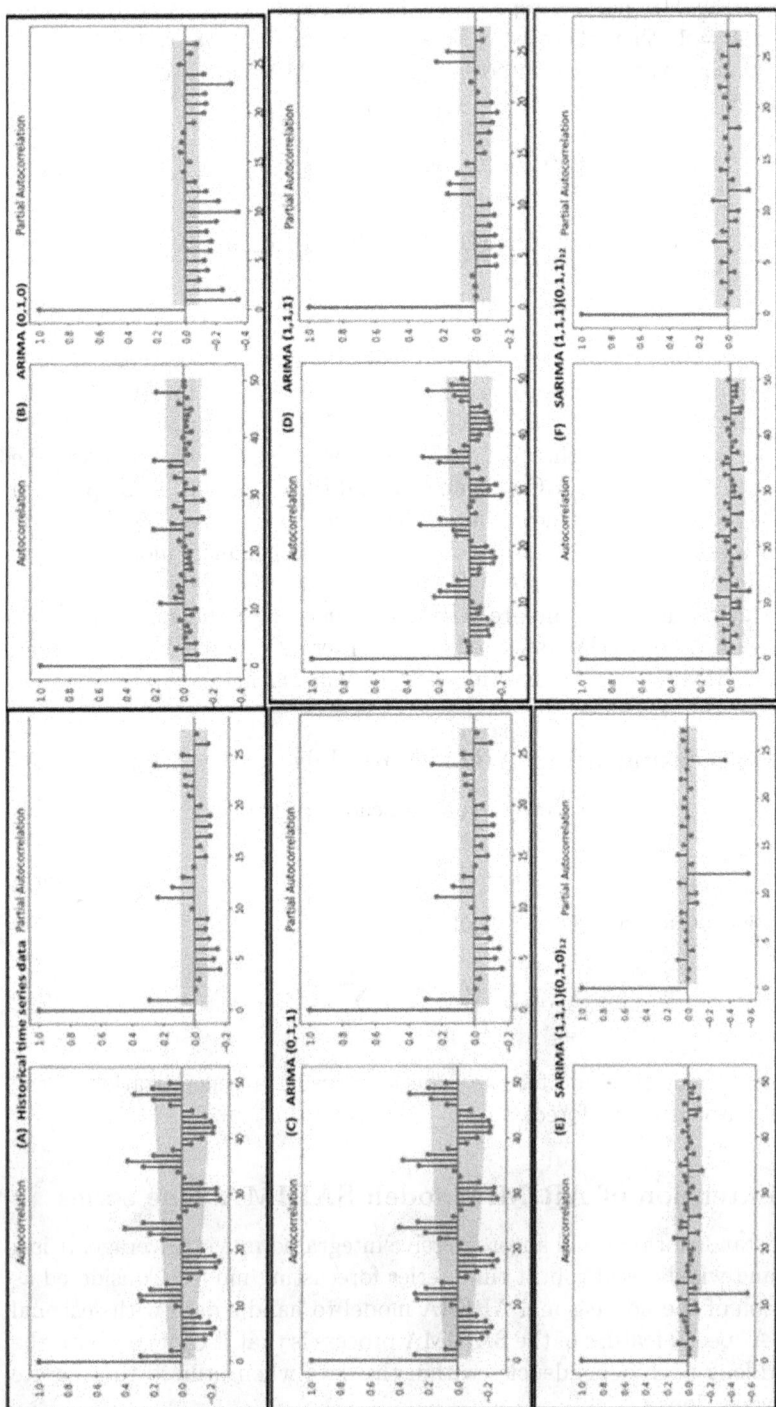

FIGURE 3.1
Images of ARIMA and SARIMA ACF and PACF.

Mathematical representation of SARIMA is as follows:

$$(1 - \phi_1 B)(1 - \Phi_1 B^s)(1 - B)(1 - B^s)y_t = (1 + \theta_1 B)(1 + \Theta_1 B^s)\epsilon_t, \quad (3.8)$$

where Φ_1 is the seasonal autoregressive coefficient, Θ_1 is the seasonal moving average coefficient, and s is the seasonal period ($s = 12$ for months, $s = 4$ for quarters, etc.).

3.3.6   Special functions of an ARIMA model

A visual illustration of ARIMA and SARIMA ACF and PACF is provided as Figure 3.1.

3.4   Chapter 3 Questions

1. What are the characteristic polynomials of an ARIMA process? Show the mathematical/statistical illustrations.

2. Write a code in R to generate an ARIMA$(0, 1, 0)$ process and its associated ACF?

3. What kind of decay is visible in an ACF plot of an ARIMA process?

4. What kind of decay is visible in an PACF plot of an ARIMA process?

5. Write a code in R to generate an ARIMA$(0, 0, 0)$ process and its associated ACF?

6. Write a code in R to generate an ARIMA$(0, 1, 2)$ process and its associated ACF?

7. Write a code in R to create an ARIMA$(0, 1, 1)$ process and its associated PACF?

8. Write a code in R to generate a visual plot of the spectral density linked with the ARIMA process in question 6 above?

9. Write a code in R to generate a visual plot of the spectral density linked with the ARIMA process in question 7 above?

10. Write a code in R to generate an ARIMA$(0, 2, 2)$ process and its associated ACF?

11. What kind of decay is visible in an ACF plot of a SARIMA process?

12. What kind of decay is visible in an PACF plot of a SARIMA process?

13. Write the expansion of the seasonal filter found in a SARIMA process?

14. List the positive characteristics of a SARIMA process?

4

Fractional Differencing and ARFIMA Process with White Noise

Synopsis: In this chapter, the focus will be on a variation of ARIMA models – long memory models (fractional differences). After successfully comprehending the material in this chapter, the reader will be able to:

1. Identify and interpret simple fractionally differenced time series models.

2. Decide when to take first differences versus fractional differences.

3. Identify and interpret ARFIMA models.

4. Establish the links and overall evolution of ARFIMA models towards generalized families of time series models.

4.1 Introduction

The stochastic analysis of time series began with the introduction of autoregressive moving-average (ARMA) model by Whittle (1951), later popularization of it by Box and Jenkins (1970) and the subsequent development of a number of path breaking research endeavors. In particular in the early 1980s, the introduction of long memory processes became an extensive practice among time series specialists and econometricians. In their papers, Granger and Joyeux (1980) and Hosking (1981) proposed the class of fractionally integrated autoregressive moving-average (ARFIMA or FARIMA) processes,

DOI: 10.1201/9781032627007-4

extending the traditional autoregressive integrated moving-average (ARIMA) series with a fractional degree of differencing. A hyperbolic decay of the auto-correlation function (acf) and an unbounded spectral density peak at or near the origin are two special characteristics of the ARFIMA family in contrast to exponential decay of the acf and a bounded spectrum at the origin in the traditional ARMA family. In addition to the mle (maximum likelihood) approach, Geweke and Porter-Hudak (1983) have considered the estimation of parameters of ARFIMA using the frequency domain approach. Chen et al. (1994), Reisen (1994), and Reisen et al. (2001) have considered the estimation of ARFIMA parameters using the smoothed periodogram approach. Additional expositions presented in Andel (1986), Brockwell and Davis (1991), Beran (1994), Rangarajan and Ding (2003), Chan and Palma (2006), Teyssiere and Kirman (2007), Giraitis et al. (2012), Beran et al. (2013) and references theirin provide a comprehensive discussion about long memory series estimation.

Utilizing fractional differencing of Hosking (1981), Gray et al. (1989) developed another class of time series known as Gegenbauer ARMA abbreviated as GARMA using the theory of Gegenbauer polynomials. Such a generalized class can be used to represent long memory depicting multiple unbounded spectral peaks away from the origin unlike in the ARFIMA case.

4.2   Fractionally Differenced Long Memory Processes

Section 5.1 of Shumway and Stoffer (2016) gives a brief overview of "long memory ARMA" models. This type of model may possibly be used when the ACF of the series tapers slowly to 0.

The usual solution in this situation is to explore the first differences of the series. Often, data for which a first difference is successful will typically have a first lag autocorrelation quite close to 1.

Note! A model for a first difference could be written in the format

$$X_t - X_{t-1} = AR \quad and \quad MA \quad terms.$$

It can be rewritten as:

$$X_t = X_{t-1} + AR \quad and \quad MA \quad terms.$$

In such a formulation, we have a first lag AR type of term with a coefficient equal to 1. It creates a first-order autocorrelation for the original series close to 1.

In some instances, however, we may see a persistent pattern of nonzero correlations that begins with a first lag correlation that is not close to 1. In these cases, models that incorporate "fractional differencing" may be useful. A simple model that utilizes fractional differencing is

$$(I - B)^\delta Y_t = X_t,$$

where δ is a value such that $|\delta| < 0.5$ and X_t is the usual white noise term.

Mathematically, this model can be expanded to be an infinite-order AR with coefficients that may taper (very) slowly toward 0.

It is clear that long memory processes do not have well defined statistical properties. Fortunately, such long memory processes can be transformed to short memory processes using suitable filters. Fractional differencing filters are widely used in long memory modeling contrary to integer differencing filters in ARIMA modeling.

4.2.1   Fractional differencing

Suppose that $\{Y_t\}$ is a long memory stationary time series. It can be shown that the time series Y_t can be transformed to a short memory series X_t through a fractional filter of the form

$$X_t = (I - B)^\delta Y_t, \ \delta \in (0, 0.5).$$

See, for example, Granger and Joyeux (1980) and Hosking (1981).

In a fractionally differenced model, the difference coefficient δ is a parameter to be estimated. R package arfima can be used to do it. Again, an indication that this model might be useful is a slowly tapering sample ACF without particularly high autocorrelations.

Exercise 1:

Step 1:   Write down the equation of a simple time series using an integer filter.

Step 2:   Write down the equation in step 1 using a fractional filter instead of an integer filter.

Step 3:   Distinguish and differentiate the two filters.

Step 4:   List down the findings of Step 3.

Example 1 of this chapter looks at a fractionally differenced model for a series length of $n = 634$ (yearly) values of a geological measurement called varve. It is a sedimentary layer of sand and silt left behind by melting glaciers. A time series plot of the data is provided in Figure 4.1 of this chapter.

Due to the period of much more extreme variability, it will be feasible to analyze the logarithm plot of the varve time series, since it might stabilize the variance. The logarithm plot is provided in Figure 4.2 of this chapter.

The sample ACF of the log-transformed data shows a persistent pattern of moderately high values. Both the ACF and the PACF are visually illustrated in Figure 4.3 of this chapter.

The arfima package in R provides an estimate for the differencing fraction of $\hat{\delta} = 0.373$. Therefore, the estimated model will be $(1 - B)^{0.373} Y_t = X_t$,

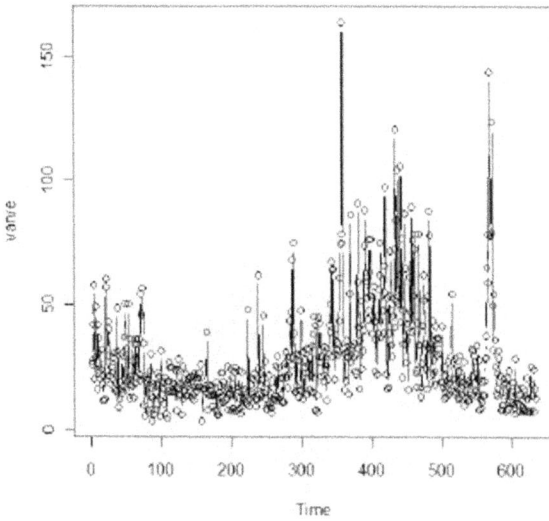

FIGURE 4.1
Varve time series realization plot.

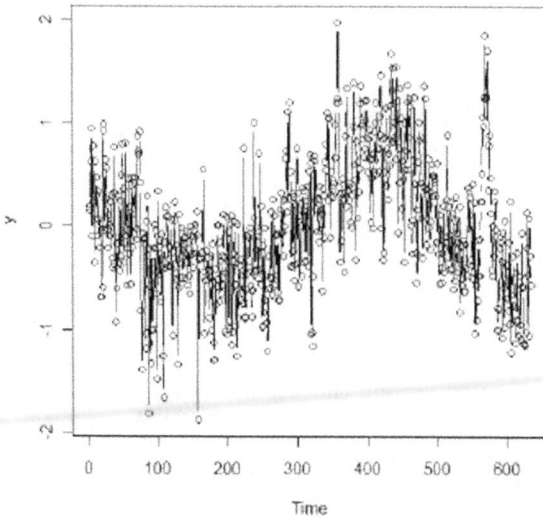

FIGURE 4.2
Plot of log-transformed varve time series.

where Y_t is the centered log-transformed log varve series. It's a model that will provide a good fit to the data as illustrated by the ACF and PACF plots of the residuals shown in Figure 4.4 of this chapter.

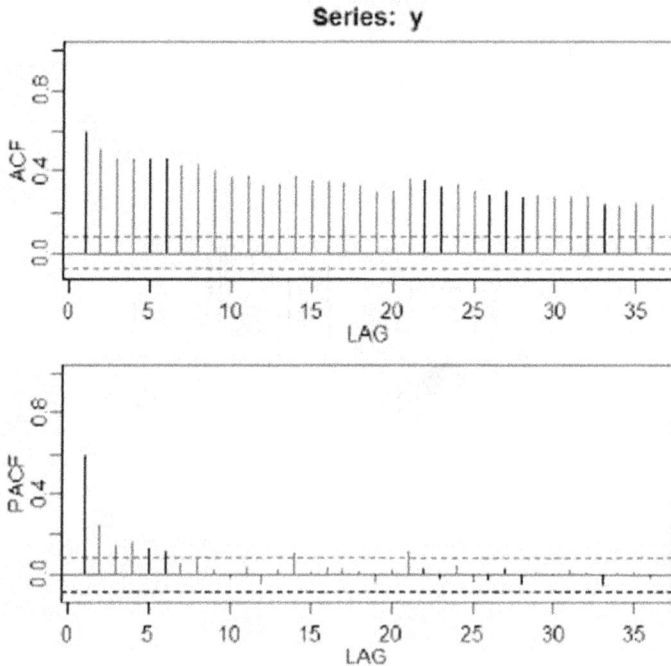

FIGURE 4.3
ACF and PACF plots of log-transformed varve time series.

The described model can be expanded to include AR and MA terms as well as the fractional difference. Such models are called ARFIMA models. To identify an ARFIMA model, we first use a simple fractional difference model $(1 - B)^\delta Y_t = X_t$ and then explore the ACF and PACF of the residuals from this model. This is analogous to exploring the ACF and PACF of the first differences when we carry out the usual steps for nonstationary data. The arfima package can be used to fit general ARFIMA models.

Note: Main difficulty is that a fractional difference is difficult to interpret. Basically, it's a mathematical device that is used to expand a model into a high-order AR with autocorrelations that match the persistent "long memory" pattern of the series ACF.

The R Code required to run the example provided in this chapter is given below. For the following code to run, need to install the arfima package; it is strongly encouraged over the fracdiff package covered in certain texts. Once the package is installed can skip the installation step and use library(arfima) to complete the analysis.

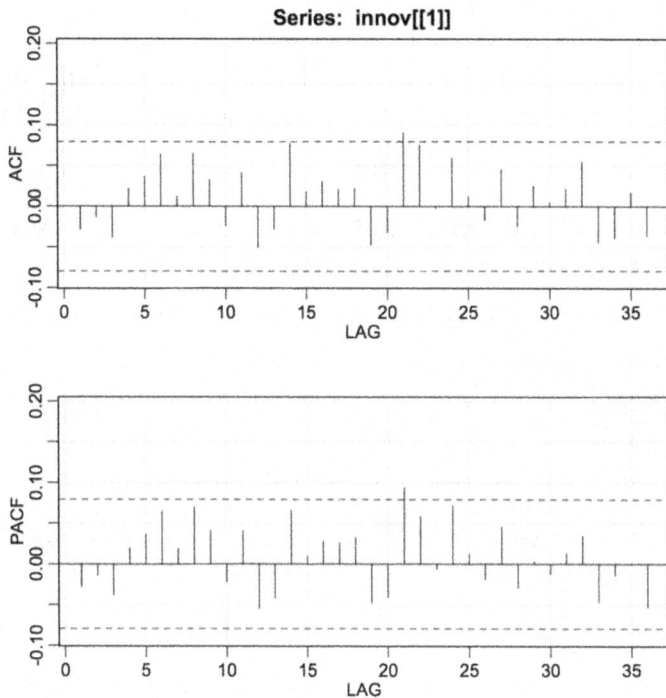

FIGURE 4.4
ACF and PACF residual plots of log-transformed varve time series.

$varve = scan(\text{``}varve.dat\text{''})$
$varve = ts(varve)$
$library(astsa)$
$install.packages(\text{``}arfima\text{''})$
library(arfima)
$y = log(varve) - mean(log(varve))$
$acf2(y)$
$Estimated:$
$varvefd = arfima(y)$
$summary(varvefd)$
$d = summary(varvefd)coef[[1]][1]$
$dprintsthevalueofdtothescreen$
$se.d = summary(varvefd)coef[[1]][1, 2]$
$se.dprintsthestandarderrorofdtothescreen$

Residuals
$resids = resid(varvefd)[[1]]$
$resid(varvefd)isalistand[[1]]accessestheresidualsasavector$

$plot.ts(resids)$

$acf2(resids)$

The above fractional differencing is used in long memory time series modeling and analysis. A much more generalized scheme of fractional differencing is presented in the next subsection.

4.2.2 Generalized fractional difference operator and its properties

This section considers generalized fractional operators.

- Consider the process satisfying

$$(I - \alpha B)^{\delta} X_t = \epsilon_t; \quad -1 < \alpha < 1; \ \delta > 0 \quad and \quad \{\epsilon_t\} \sim WN(0, \sigma^2). \quad (4.1)$$

Clearly, this covers the standard first-order autoregressive [AR(1)] family when $\delta = 1$ and when $\alpha = 1$, $0 < \delta < 1/2$, $\{X_t\}$ becomes a fractionally differenced white noise (FDWN). This is used as building blocks for long memory ARFIMA(p, d, q) time series modeling given by

$$\phi(B)(1 - B)^{\delta} X_t = \theta(B)\epsilon_t, \quad (4.2)$$

where $\phi(B)$ and $\theta(B)$ are stationary AR(p) and invertible MA(q) operators such that $\phi(B) = 1 - \phi_1 B - ... - \phi_p B^p$, $\theta(B) = 1 + \theta_1 B + ... + \theta_q B^q$ with zeros outside the unit circle.

- A second-order model is often a natural extension of (4.1) and is defined by

$$(1 - \alpha_1 B - \alpha_2 B^2)^{\delta} X_t = \epsilon_t, \quad (4.3)$$

where $1 - \alpha_1 z - \alpha_2 z^2 \neq 0$ for all $|z| \leq 1$, $\delta > 0$, and $\{\epsilon_t\} \sim WN(0, \sigma^2)$.

In (4.3), conditions $\alpha_2 + \alpha_1 < 1$, $\alpha_2 - \alpha_1 < 1$, and $-1 < \alpha_2 < 1$ need to satisfy α_1 and α_2 to ensure stationary solutions. An interesting case arises when the stationary assumption of (4.3) fails. That is at least one of the above three inequalities does not hold, and this will be considered in a future paper.

We could write the model (4.3) as:

$$[(1 - \xi_1 B)(1 - \xi_2 B)]^{\delta} X_t = \epsilon_t,$$

where $\xi_1 + \xi_2 = \alpha_1$ and $\xi_1 \xi_2 = -\alpha_2$.

It can be shown that the solution of this model is given by

$$X_t = \sum_{k=0}^{\infty} \sum_{l=0}^{\infty} \frac{\Gamma(k + \delta)\Gamma(l + \delta)\xi_1^k \xi_2^l}{\Gamma(k + 1)\Gamma(l + 1)\Gamma^2(\delta)} \epsilon_{t-k-l},$$

where $\Gamma(\cdot)$ is the gamma function satisfying $\Gamma(k) = (k-1)!$ for integer $k \geq 1$. The model in (4.3) is called the generalized second order autoregression with index δ. For simplicity, we call it generalized autoregressive of order 2 [GAR(2)] model. It is clear that the class generated by (4.3) is more flexible than the standard AR(2) model. Although the autocorrelation function of the GAR(2) model can be obtained, it is not mathematically tractable as there is no closed-form solution. The spectral density $f(\omega)$ is given by

$$f(\omega) = \frac{\sigma^2}{2\pi}[(1 + \alpha_1^2 + \alpha_2^2) - 2\alpha_1(1 - \alpha_2)\cos\omega - 2\alpha_2\cos 2\omega]^{-\delta}.$$

See Shitan and Peiris (2008) for details.

- An interesting class of models will arise when $\alpha_2 = -1$ in (4.3). This leads to Gegenbauer processes which extends a mathematically elegant class of time series models with very useful applications as given below.

Exercise 2: Provide the theoretical expansion of the polynomial $(1 - B)^{\delta}$ using the binomial theorem?

4.3   Chapter 4 Questions

1. What criteria are used to distinguish between "integer differencing (introduced in the previous chapter)" and "fractional differencing"?

2. How does an ARFIMA model change from an ARIMA model?

3. Explain the evolution beginning with a short memory ARMA model to a long memory ARFIMA model?

4. Why is fractional differencing important in time series analysis?

5. Compare and contrast "fractional differencing" with "integer differencing"?

6. What role does the ACF and the PACF play in the fractional differencing process of a time series?

7. How does the ACF and PACF of residual plots used to assess a time series in the post fractional differencing phase?

8. Provide the output after running the R code provided in the chapter?

9. Theoretically how does a fractionally differenced ARFIMA model transform itself to a GAR(2) model?

10. What variation in the time or frequency domain leads GAR(2) models toward converting into Gegenbauer processes as yet another mathematically elegant class of time series models?

11. What does a hyperbolic decay of the ACF and an unbounded spectral peak at or near the origin of an ARFIMA model symbolize in terms of characteristics of the time series model?

5

Short, Intermediate, and Long Memory Properties of Time Series

Synopsis: All types of time series could be segregated based on serial correlation between the lagged observations. In a sense it will be an evaluation of a specific observation and its preceding value(s). Based on the duration and the strength of the serial correlation within the series it could be classified as a short, intermediate, or a long memory series.

Certain special functions of a time series such as acf, pacf, and sdf will become the instruments to identify the memory type of a time series. Table 5.1 given below would provide a summary in terms of the characteristics of memory types.

5.1 Background

Note: Graphical illustrations of the acfs and pacfs of short and long memory processes are provided below as Figures 5.1 and 5.2. Guegan (2005) provides a number of alternative characteristics of long memory processes. Interestingly, Wang et al. (2006) introduced a characteristic-based clustering method to capture the characteristic of long-range dependence (self-similarity).

Note: Processes in which the decay of ρ_k takes a shape in between exponential and hyperbolic arcs depict *intermediate memory*.

Remark 5.1. Also, pacf of each memory type will provide corresponding shapes related to that of the acf.

Exercise 1: Sketch the acf and the pacf of an intermediate memory process?

DOI: 10.1201/9781032627007-5

TABLE 5.1
Characteristics of memory types

Short Memory Series	Long Memory Series
Stationary	Stationary
Exponential decay of ρ_k	Hyperbolic decay of ρ_k
$\rho_k \sim a^k$ for $|a| < 1$	$\rho_k \sim k^{-d}$ for $d > 0$
$\sum |\rho_k| < \infty$	$\sum |\rho_k| = \infty$
$lim_{\omega->0} f(\omega)$ exists & bounded	$lim_{\omega->0} f(\omega)$ does not exist
	or $f(\omega)$ unbounded

5.2   Evolution of Long Memory Time Series

The stochastic analysis of time series began with the introduction of ARMA model by Whittle (1951), popularization by Box and Jenkins (1970) and subsequent developments of a number of path breaking research endeavors. It was a result of combining AR and MA time series. In particular, in the early 1980s, the introduction of long memory processes became an extensive practice among time series specialists and econometricians. In their papers, Granger and Joyeux (1980) and Hosking (1981) proposed the class of fractionally integrated autoregressive moving average (ARFIMA or FARIMA) processes, extending the traditional autoregressive integrated moving average (ARIMA) series with a fractional degree of differencing. A hyperbolic decay of the acf, and an unbounded spectral density peak at or near the origin are two special characteristics of the ARFIMA family in contrast to exponential decay of the acf and a bounded spectrum at the origin in the traditional ARMA family. In addition to the mle (maximum likelihood estimation) approach, Geweke and Porter-Hudak (1983) have considered the estimation of parameters of ARFIMA using the frequency domain approach. Chen et al. (1994), Reisen (1994), and Reisen et al. (2001) have considered the estimation of ARFIMA parameters using the smoothed periodogram approach. Additional expositions presented in Andel (1986), Brockwell and Davis (1991), Beran (1994), Rangarajan and Ding (2003), Chan and Palma (2006), Teyssiere and Kirman (2007), Giraitis et al. (2012), Beran et al. (2013), and references provide a comprehensive discussion about long memory series estimation.

Utilizing fractional differencing of Hosking (1981), Gray et al. (1989) developed another class of time series known as Gegenbauer ARMA abbreviated as GARMA using the theory of Gegenbauer polynomials. This generalized class can be used to represent long memory depicting multiple unbounded spectral peaks away from the origin unlike in the ARFIMA case of Hosking (1981) at the origin (Figures 5.3 and 5.4 provide a visual illustration). A detailed

ARMA: Autocorrelation (left) and Partial Autocorrelation (right)

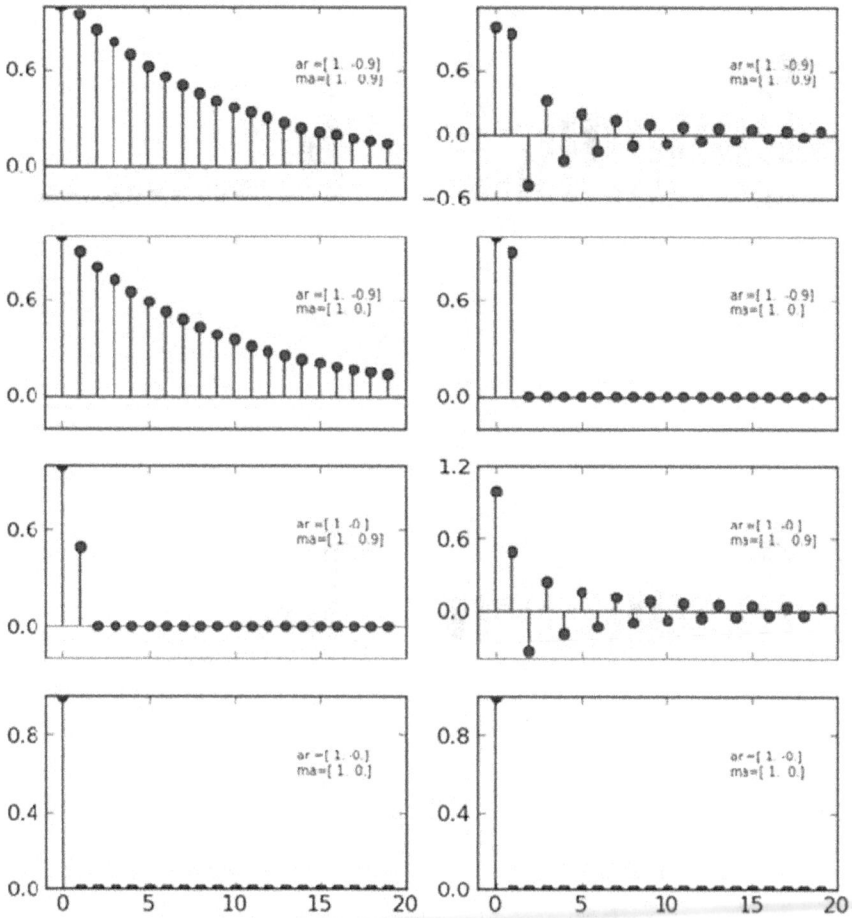

FIGURE 5.1
Autocorrelation function of standard short memory process.

analysis of the long memory features of GARMA time series is illustrated through the unbounded spectral density around the Gegenbauer frequency by Chung (1996). The existing literature suggests that the ARFIMA model of Hosking (1981), and the generalized fractional GARMA model of Gray et al. (1989) have unique time series characteristics of their own. In yet another development, fractionally differenced long memory model parameters were estimated using maximum likelihood and least squares with their convergence

Telegraph Road

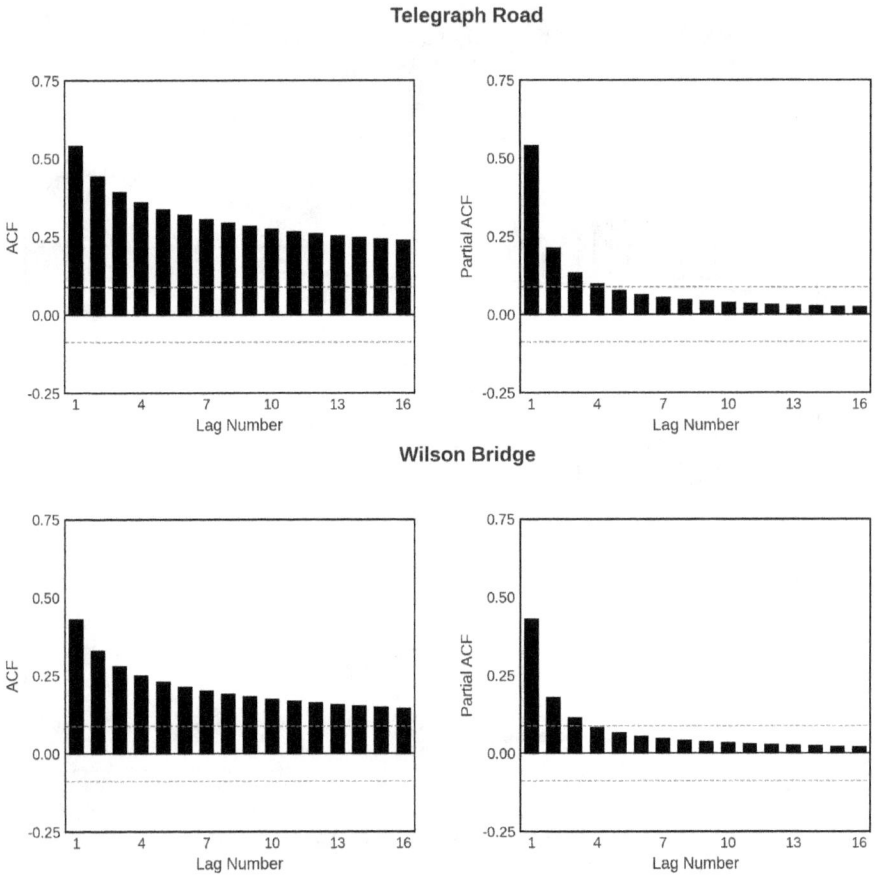

Wilson Bridge

FIGURE 5.2
Autocorrelation function of standard long memory Process.

rates, limiting distribution and strong consistency by Yajima (1985). However, the utilization of state space methodology following the work of Anderson and Moore (1979), Pearlman (1980), Harvey (1989), Aoki (1990), Brockwell and Davis (1996), Chan and Palma (1998), Durbin and Koopman (2001), Harvey and Proietti (2005), Palma (2007), and Grassi and De Magistris (2014) in estimating parameters, establishing predictive accuracy and an optimal lag order is not evident in the literature.

Therefore, state space modeling coupled with Kalman filter (KF) is deemed suitable as an alternative technique to estimate and forecast GARMA time series. In developing such methodology assessing a generic representation becomes a useful topic. It is presented in the next subsection.

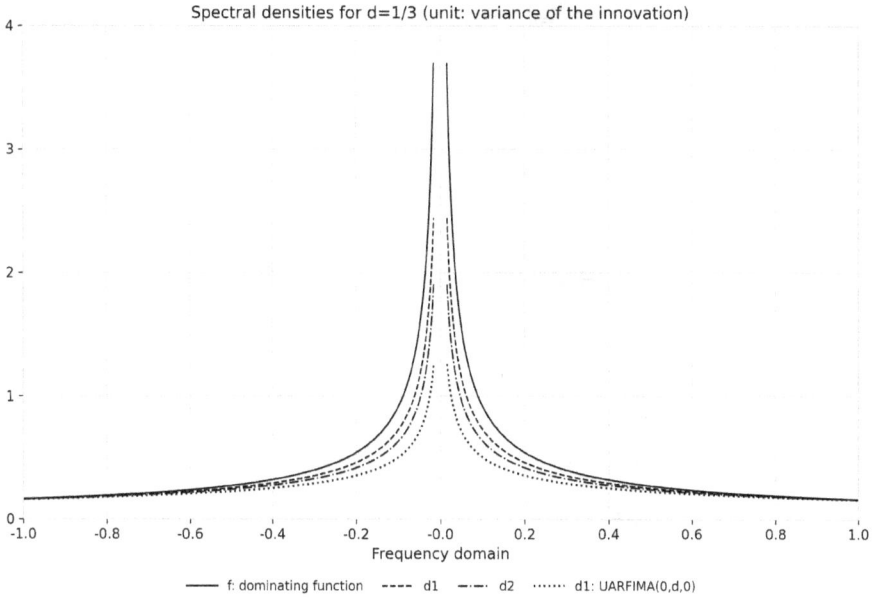

FIGURE 5.3
Spectral density of standard long memory process with peak at the origin.

5.3 State Space Representation of a Generic Time Series – An Overview

In the literature, there are various state space configurations proposed by different authors. Each of them consists of two fundamental equations for a process $\{X_t\}$. Suppose that an observed vector series $\{X_t\}$ can be written in terms of an observed state vector $\{\alpha_t\}$ (of dimension $m \times 1$). This first equation is known as the *observation (measurement) equation* and is given by

$$X_t = Z_t \alpha_t + \epsilon_t, \quad t = 1, 2, ..., \tag{5.1}$$

where $\{\epsilon_t\} \sim WN(0, \sigma_{\epsilon^2})$ and $\{Z_t\}$ is a sequence of $1 \times m$ matrices with m being the lag cut off point.

The second equation known as *state (transition) equation* determines the evolution of state α_t at time "t" in terms of the previous state α_{t-1} and the noise term ϵ_t. It is given by

$$\alpha_t = T_t \alpha_{t-1} + \epsilon_t, \quad t = 2, 3, ..., \tag{5.2}$$

where T_t is a sequence of $m \times m$ matrices and $\{\epsilon_t\} \sim WN(0, \sigma_\epsilon^2)$.

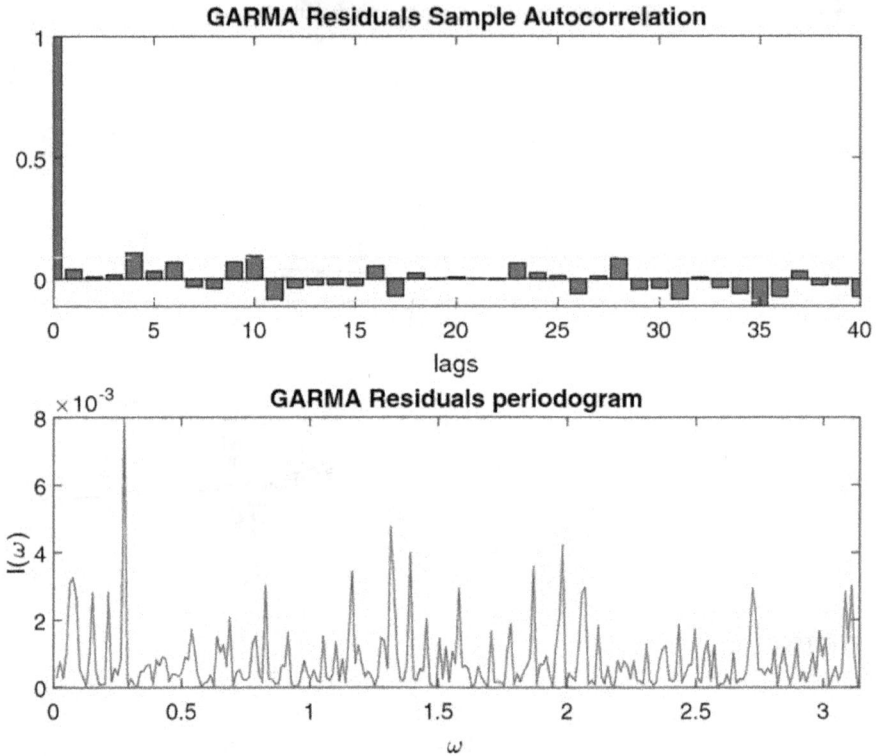

FIGURE 5.4
Spectral density of gegenbauer long memory process with peak away from the origin.

It is assumed that the initial state α_1 is uncorrelated with all the noise terms. In general,

$$\alpha_t = f_t(\alpha_1, \epsilon_1, ..., \epsilon_{t-1})$$

and

$$X_t = g_t(\alpha_1, \epsilon_1, ..., \epsilon_{t-1}, \epsilon_t)$$

Note: The state space representation given by (5.1) and (5.2) is not unique, but useful in developing important results in linear and nonlinear time series models utilizing *recursive estimation methods*. It becomes the topic of discussion in the next subsection.

5.4 Recursive Estimation

Recursive estimation revolves around the process known as *Recursion*. In a mathematical and computational sense, it is defined as the mechanism of a function calling itself within an embedded structure. Estimators of this type are aimed at tracking time varying parameters. Therefore, it is desirable to make calculations recursively to save computation time. In recursive estimation a parameter estimate defined as P_t (a vector of the parameter at time t) could be obtained as a function of the previous estimate P_{t-1} and of the current (or new) measurements.

Some standard recursive algorithms are utilized in practice to perform the tracking of time series parameters through estimation and forecasting. Of them, the most popular algorithm utilized by state space modeling specialists for estimation and prediction is known as the KF and becomes the focal point of the next subsection.

5.5 KF and Estimation Process

A set of recursions known as the KF was introduced by Kalman (1961) and Kalman and Bucy (1961) to provide estimates of parameters in a state space model of a time series or a linear dynamic system disturbed by Gaussian white noise. Approximate maximum likelihood estimation and prediction of time series parameters can be executed by adopting a state space approach coupled with the KF.

Due to the presence of stochastic elements in the system, it uses a series of measurements observed over time containing random variations (noise) to return innovations of the model in creating the log-likelihood and profile likelihood functions. This gives the optimal maximum likelihood estimates (MLE's) of the model parameters through the *Kalman gain* (which shows the effect of estimate of the previous state to the estimate of the current state), with a minimal prediction error variance. Furthermore, a discussion about the KF formulation of the likelihood function can be found in Jones (1980). The methodology presented in this section could probably be employed to model a seasonal GARMA series and becomes yet another discussion topic in a subsequent chapter of this book. The same methodology could be employed to analyze seasonality in long memory Gegenbauer series as an extension to address the prevalent issue.

It will be important in terms of modeling another extended class of Gegenbauer processes since in practice, researchers have noticed that many time series consist of seasonality in both ARFIMA and GARMA with long memory. This phenomenon had been observed in monetary aggregates of Porter-Hudak

(1990), revenue series of Ray (1993), inflation rates of Hassler and Wolters (1995), quarterly gross national product and shipping data of Ooms (1995) and monthly Nile river flows of Montanari et al. (2000). Several statistical modeling methodologies for such processes have been developed. The fractional Gaussian noise process of Abrahams and Dempster (1979), seasonal fractionally integrated autoregressive moving average (SARFIMA) model of Porter-Hudak (1990), flexible seasonal fractionally integrated process (flexible ARFISMA) of Hassler (1994), k-GARMA process of Woodward et al. (1998), seasonal long range-dependent process of Palma and Chan (2005), seasonal fractionally integrated process of Reisen et al. (2006), and the KF-based SARFIMA process of Bisognin and Lopes (2009) are some examples available in the literature. Properties of these models have been investigated by Giraitis and Leipus (1995), Chung (1996), Arteche and Robinson (2000), Velasco and Robinson (2000), Giraitis et al. (2001), Palma (2007), Arteche (2007), Koopman et al. (2007), Bisognin and Lopes (2009), Hsu and Tsai (2009), and Arteche (2012).

Similarly, in many financial time series modeling problems, it is known that heteroscedasticity plays an important role. Such models in common use are the autoregressive conditionally heteroskedastic (ARCH) model of Engle (1982) and its generalized ARCH (GARCH) model due to Bollerslev (1986). An extension of it by Baillie, et al (1996a) resulted in a fractionally integrated GARCH (FI-GARCH) to model the conditional variance. Incorporating heteroscedasticity in ARFIMA models with GARCH errors has been studied by Ling and Li (1997). An attempt to capture and blend some of these established GARCH features in introducing conceptual properties of a new class of models with conditionally heteroskedastic errors is presented in Dissanayake and Peiris (2011). In such a context, it could be based on the generalized fractional operator that was used by Anh et al. (1999) and later developed by Peiris (2003), and Shitan and Peiris (2008, 2013). This scaled down new operator with further applications was employed by Peiris et al. (2005) and Peiris and Thavaneswaran (2007) in long memory models driven by heteroskedastic GARCH errors. Employing the state space methodology to estimate the new class of models with GARCH errors could be found in Dissanayake, et al. (2014b).

Several studies in the literature report evidence of long memory in empirical volatility returns as illustrated by Robinson (1991), Shephard (1996), Lobato and Savin (1998), and Baillie (1996). McAleer and Medeiros (2008) proposed a flexible model to describe nonlinearities and long memory in time series dynamics for purposes of forecasting volatility. Additionally, Lieberman and Phillips (2008) have provided some analytical explanations for long memory behavior that has been observed in realized volatility. Furthermore, a simple additive cascade approximate long memory model of realized volatility was proposed by Corsi (2009). In a recent development, Rossi and Santucci de Magistris (2014) illustrated if instantaneous volatility is a long memory process, then the integrated variance is characterized by the same long-range dependence. Furthermore, Bos et al. (2014) provide evidence that properties of time series such as US inflation are unstable over time through the model-based analysis of a specified long memory ARFIMA process with variance modeled by stochastic volatility. Most econometric models dealing with long memory and heteroskedastic behavior are nonlinear in the sense that the noise sequence is not necessarily independent. Such models based on conditional variance are primarily driven by a GARCH process.

Furthermore, it has become a customary practice in applied time series analysis to conduct tests to assess whether a time series is stationary or to be integrated at a suitable degree of differencing. Such testing to check if a variable is integrated of order one or stationary without being differenced using null hypotheses was illustrated by Phillips and Xiao (1998). This procedure became very popular among econometricians and was made famous under the theme of *"unit root hypothesis"* (see, for example, Phillips and Xiao (1998), Dolado et al. (2002)). Such hypotheses of stationary series with respect to unit roots could be extended to fractional processes with long memory. A unit root test for fractionally integrated processes has been proposed in Dolado et al. (2002) and asymptotic results of a similar test is presented in Wang et al. (2003). Additionally, Taylor (2005) introduced a set of new tests to assess constant trend stationarity against the change in persistence from trend stationarity to difference stationarity or vice versa. Furthermore, Ohanissian et al. (2008) proposed a statistical test to distinguish between true and spurious long memory.

Hypothesis testing of linear or nonlinear constraints on the parameters of econometric models could be performed using one of three methods. They are "Likelihood Ratio (LR), Wald and Lagrange Multiplier (LM) tests. All are asymptotically equivalent, but in most finite samples the results will differ. In many ways, the LR test is conceptually the easiest to use and can be employed for any model estimated by maximum or quasi-maximum likelihood. Finding an appropriate test for a specific time series generally becomes a challenge. For a long memory fractionally differenced Gegenbauer process and related parameter estimation through state space modeling and KF the estimation process employs a quasi-maximum-likelihood function. Therefore, an LR test becomes the most feasible assessing tool in such a context. Generating results of such a test with an acceptable power will be another added feature in this book.

5.6   Chapter 5 Questions

1. Describe what is meant by "memory" in time series analysis.

2. What type of functions are utilized to distinguish memory types?

3. Describe the characteristics of memory types.

4. What are the memory types?

5. Name a special type of a modeling concept that could be deployed to model long memory processes?

6. To employ state space modeling of long memory processes what features are required in the time series to be evaluated?

7. Define Kalman filter.

8. Name a time series that you are aware of that could be an example of a short memory model.

9. Name a time series that you are aware of that could be an example of a long memory model.

10. How would you distinguish and differentiate between the concepts of a "Standard long memory process" and a "Gegenbauer long memory process"?

6

Standard Long Memory and State Space Modeling of ARFIMA Process with White Noise

Synopsis: In this chapter, the focus will be on a variation of ARIMA models – long memory models (fractional differences). After successfully comprehending the material in this chapter, the reader will be able to:

1. Identify and interpret simple fractionally differenced time series models.

2. Decide when to take first differences versus fractional differences.

3. Identify and interpret ARFIMA models.

4. Establish the links and overall evolution of ARFIMA model toward generalized families of time series models.

6.1 Introduction

Long memory data arise in a wide variety of subject areas, from hydrology as well as epidemiology to economics and econometrics. Such time series were brought to light by Hurst (1951) in introducing hydrological time series with attributes of long-range dependence that subsequently received unparallel attention in the literature. Use of fractional processes in economics and econometrics was introduced in a seminal paper by Granger (1980) and Granger and Joyeux (1980). Refer volumes by Beran (1994) and Palma (2007) and collections of Robinson (2003) and the included references. Initial point of econometric literature on autoregressive fractionally integrated moving average (ARFIMA) models had been motivated due to many economic and financial time series showing evidence of being neither I(0) nor I(1). At present, a broad range of applications in finance and macroeconomics show that fractional integration and long memory are relevant, as examples refer Diebold et al. (1991) for exchange rate data, Andersen et al. (2001a) and Andersen et al. (2001b) for financial volatility processes, and Baillie et al. (1996) for inflation data. Initial articles on estimation of long-range dependent models could be found in Fox and Taqqu (1986), Dahlhaus (1989), Sowell (1992), and Robinson (1995). Refer Chan and Palma (2006) for a substantial review of the relevant material.

As an alternative configuration, Chan and Palma (1998) proposed a state space approach to compute the maximum likelihood (referred to as ML hereinafter) estimates for an ARFIMA model. The authors proposed to truncate the infinite MA and AR representation of an ARFIMA model and establish the ARFIMA by casting it in state space. Long memory parameter, d, can be then estimated by means of the Kalman filter (referred to as KF hereinafter). Estimates obtained by this method were consistent, asymptotically normal and efficient under mild regularity conditions. The introduced methodology, although conceptually simple, was computationally tedious and cumbersome during the last decade of the last millennium, and was not commonly used. However, with the novel computational capabilities at hand nowadays, it is possible to estimate such models in a few seconds even for large datasets. Multiple simulation studies, by Rea et al. (2008), Nielsen and Frederiksen (2005), and Haldrup and Nielsen (2007) provide a comparative assessment between different estimation strategies, without considering the state space alternative. It emphasizes on the fact that, although, great effort has been spent in the estimation of fractional processes with semiparametric and maximum likelihood methods, little has been done to explore the state space option. In this chapter, an extensive Monte Carlo simulation exercise will be presented in the long memory framework.

Practical importance of state space methods relies on the possibility of a straightforward modeling approach for measurement errors, outliers, level shifts, and missing values. It can be illustrated through simulations that the KF provides unbiased, and efficient estimates of the model parameters also in such cases. Proposed and presented Monte Carlo simulations of this chapter are therefore intended as an attempt to explore the potentiality of the state space approach within the long memory framework. In such a context, consideration is given to several parametric and semiparametric estimation methods for ARFIMA models, and compare them with the state space alternative. By taking into consideration the work done by Nielsen and Frederiksen (2005) and Haldrup and Nielsen (2007), the bias and the root mean squared error (RMSE) are adopted to measure the finite sample performances of the estimators.

Results of this chapter can be summarized as follows: firstly, state space methodology is a valid alternative to the usual estimation procedures and robust to non-Gaussian shocks and overspecification. In case of under-specification, the Akaike information criterion, AIC, always selects the right model. Secondly, when series at hand has missing observations, the state space estimation method is superior to other traditional techniques as it has low bias and RMSE. Third characteristic is that, in case of measurement errors, the KF largely outperforms the traditional estimators as well as the corrected local Whittle method presented in Hurvich et al. (2005).

6.2 ARFIMA Process

ARFIMA(p, d, q) process y^t is defined as:

$$\Phi(B)(1 - B)^d y_t = \Theta(B)\eta_t, \tag{6.1}$$

where (η_t) is a sequence of independent random variables with a constant mean and variance equal to σ_η^2, backshift operator B is such that $B(y_t) = y_{t-1}$, $\Phi(B) = 1 - \phi_1 B - \ldots - \phi_p B^p$ is the AR polynomial, $\Theta(B) = 1 - \theta_1 B - \ldots - \theta_q B^q$ is the MA polynomial, and $(1 - B)^d$ is the fractional difference operator.

Parameter d determines long memory of the process. If $d > -0.5$, the process is invertible and possesses linearity. On the contrary, if $d < 0.5$, it becomes covariance stationary. Furthermore, if $d > 0$ the process will have long memory, since the autocorrelations will taper out at a hyperbolic rate (will not be absolutely summable) in comparison with much faster exponential rates in the weak dependence scenario.

For $d \in (0, 0.5)$ model (6.1) becomes a stationary long-memory process with nonsummable autocorrelations, defined by $\sum_{k=0}^{\infty} |\rho_k| = \infty$.

If $d = 0$, the spectral density is bounded at the origin and the process has only weak dependence (short memory), and it becomes the well-known ARMA process. In most practical cases, the parameter $d \in (0, 0.5)$ has been proved to be relevant for many applications. The fractional difference operator $\Delta^d = (1 - B)^d$ in equation (6.1) is defined by its binomial expansion given as:

$$(1 - B)^d = \sum_{j=0}^{\infty} \frac{\Gamma(j - d)}{\Gamma(j + 1)\Gamma(-d)} B^j, \tag{6.2}$$

where $\Gamma(.)$ is the Gamma function. Hosking (1981) did show that a stationary ARFIMA(p, d, q) could admit infinite MA and AR expansions as shown below:

$$y_t = \sum_{j=0}^{\infty} \psi_j \eta_{t-j}, \tag{6.3}$$

$$y_t = \sum_{j=1}^{\infty} \pi_j y_{t-j} + \eta_t \tag{6.4}$$

Hosking (1981) also provided a formula to compute weights ψ_j and π_j for low-order ARFIMA processes. An alternative, although not equivalent, definition of long-range dependence can be stated in the frequency domain. In particular, the spectral density of the ARFIMA(p, d, q) process (6.1) can be represented as follows:

$$f_\Theta(\lambda) \sim G|\lambda|^{-2d} \quad for \quad \lambda \to 0, \tag{6.5}$$

where $\Theta = (d, \phi_1, ..., \phi_p, \theta_1, ..., \theta_q)$ and G corresponds to the spectral density of an ARMA(p, q) process, see Beran (1994).

6.3 State Space Form for ARFIMA Long Memory Processes

Long-range-dependent processes have an infinite-dimensional state space representation, with Chan and Palma (1998) proving that the likelihood of an ARFIMA process can be computed, by means of the KF, in a finite number of steps. For example, the ARFIMA(p, d, q) series has linear MA or AR representations given by formulae (6.3) and (6.4). To make the KF recursions feasible, Chan and Palma (1998) consider an approximation of equations (6.3) and (6.4) based on a truncation up to lag m. In this chapter, both types of representation will be considered. Finally, Chan and Palma (1998) provided asymptotic properties of these approximate maximum likelihood estimates. Interesting results are that under mild regularity conditions, the maximum likelihood estimators are both consistent and efficient. It is worth taking into consideration the suggestion of Palma (2007) to exploit the MA approximation in first differences that guarantees a computational advantage.

6.4 State Space Form

The state space representation consists of two equations. The first is the measurement (observation) equation, which relates the time series y_t to the state vector:

$$y_t = Z\alpha_t + D\epsilon_t, \quad t = 1, 2, ..., T, \quad \epsilon_t \sim NID(0, \sigma_\epsilon^2), \tag{6.6}$$

where Z is a $1 \times m$ matrix. The second term is the transition (state) equation that defines the evolution of the state vector α_t as a first-order vector autoregression:

$$\alpha_t = T\alpha_t + H\eta_t, \quad \eta_t \sim NID(0, Q), \tag{6.7}$$

where T is a $m \times m$ matrix and H is a $m \times g$ selection matrix, while η_t is a $g \times 1$ disturbance vector. Q is an $m \times m$ matrix of zeros, whose first element is equal to σ_η^2.

Chan and Palma (1998) and Palma (2007) proposed that there are two ways to cast a long memory model in state space form. The first is based on the AR(∞) representation, and the second is based on the MA(∞) representation. Selecting a truncation lag m that is large enough, it is sufficient for the evaluation of the quasi-likelihood. This leads to an AR(m) and MA(m) approximations that can be cast in state space form and estimate using the KF recursions, refer Harvey (1989) and Harvey and Proietti (2005) for further information.

The AR(m) approximation (SS-AR henceforth) can be written as follows:

$$Z = [1, 0, \ldots, 0],$$

$$T = \begin{bmatrix} \pi_1 & \pi_2 & \cdots & \pi_m \\ I_{m-1} & & & 0 \end{bmatrix},$$

$$H = [1, 0, \ldots, 0],$$

$$D = 0,$$

$$t = 1, 2, \ldots, T,$$

where π_j for $j = 1, \ldots, m$ comes from (6.4), refer Hosking (1981) for further details.

Corresponding MA(m) approximation (SS-MA henceforth) can be written as follows:

$$Z = [1, 0, \ldots, 0],$$

$$T = \begin{bmatrix} 0 & I_m \\ 0 & 0 \end{bmatrix},$$

$$H = [1, \psi_1, \psi_2, \ldots, \psi_m],$$

$$D = 0,$$

$$t = 1, 2, \ldots, T,$$

where ψ_j for $j = 1, \ldots, m$ comes from (6.3), refer Hosking (1981) for further information.

Note: MA and AR approximations would lead toward quasi-maximum-likelihood estimates of parameters, since an approximate (tapered) likelihood function with a high degree of accuracy would be employed with an unknown underlying statistical distribution.

TABLE 6.1

QMLE results due to the MA approximation

m	29	30	31	32	33	34	35
$\hat{d}$	0.2285	0.2288	0.2304	0.2317	0.2320	0.2376	0.2362
$\hat{\sigma}$	0.9664	0.9625	0.9622	0.9650	0.9649	0.9571	0.9660
Model bias	−0.0051	−0.0087	−0.0074	−0.0033	−0.0031	−0.0053	0.0022

TABLE 6.2

QMLE results due to AR approximation

m	9	10	11	12	13
$\hat{d}$	0.2238	0.2223	0.2215	0.2222	0.2243
$\hat{\sigma}$	0.9638	0.9633	0.9643	0.9657	0.9662
Model bias	−0.0124	−0.0144	−0.0142	−0.0121	−0.0095

6.5 Kalman Filter

The KF (see Harvey, 1989, and Durbin and Koopman, 2001) is a fundamental algorithm for the statistical treatment of a state space model. Under the Gaussian assumption, it produces the minimum mean square estimator of the state vector along with its mean square error matrix, conditional on past information; this is used to build the one-step-ahead predictor of y_t and its mean square error matrix. Due to the independence of the one-step-ahead prediction errors, the likelihood can be evaluated via the prediction error decomposition, see Schweppe (1965).

KF is nothing but a set of recursions. In the next chapter (Chapter 7) of this book, KF will be illustrated in detail for the generalized GARMA state space representation. Since the GARMA family of time series models includes the ARFIMA process as a special member, the KF recursions and illustrations of the next chapter would be applicable to ARFIMA state space configurations too.

Tables 6.1 and 6.2 provide QMLE results through MA and AR approximations due to state space modeling of an ARFIMA$(0, d, 0)$ series.

6.6 Empirical Evidence

Nile river outflow and ACPI data have been premier real data sets utilized by time series econometricians and statisticians over the years due to their

TABLE 6.3
QMLE results for Nile River data

Method	$\hat{d}$ (Standard Error)
MA Approximation	0.291 (0.012)
AR Approximation	0.278 (0.019)

TABLE 6.4
QMLE results for ACPI data

Method	$\hat{d}$ (Standard Error)
MA Approximation	0.331 (0.009)
AR Approximation	0.319 (0.015)

close relationship with standard long memory. The chosen data sets have a significant impact in econometrics, since an assessment of the Nile river outflow is important to irrigation and agricultural production yields that affect the economies of many third world African nations, while an ACPI benchmark affects the economic stability of a developed first-world country. The long memory feature becomes evident from the spectral density function (sdf) plots of the datasets with infinite peaks close to the origin. Furthermore, the autocorrelation function (acf) and partial autocorrelation function (pacf) plots depict long memory through hyperbolically decaying arcs.

In lieu of this, Nile river data from 1870 to 2011 and ACPI data from the third quarter of 1948 to the second quarter of 2015 were considered and fitted to the hybrid ARFIMA$(0, d, 0)$ state space model discussed in this paper. The datasets were downloaded from `https://datamarket.com` and `http://www.abs.gov.au` websites, and the corresponding results are provided in Tables 6.3 and 6.4.

Note: The values in brackets adjacent to the parameter estimates of d are the standard errors. From the estimate values, it is evident that the long memory property is preserved since, with both approximations of the applications, $0 < \hat{d} < 0.5$.

6.7 Useful R Codes for ARFIMA Modeling

The **arima** function given below simulates a long memory ARIMA time series, with either fractionally differenced white noise (FDWN), fractional Gaussian noise (FGN), power-law autocovariance (PLA) noise, or short memory noise and with seasonality. arfima.sim(

n,
model = list(phi = numeric(0), theta = numeric(0), dint = 0, dfrac = nu-
meric(0), H =
numeric(0), alpha = numeric(0), seasonal = list(phi = numeric(0), theta =
numeric(0),
dint = 0, period = numeric(0), dfrac = numeric(0), H = numeric(0), alpha =
numeric(0))),
useC = 3,
sigma2 = 1,
rand.gen = rnorm,
muHat = 0,
zinit = NULL,
innov = NULL,
...)

By running the following R code, a fractionally differenced ARIMA model or an ARFIMA model could be generated together with the resulting residual plot by using the R library known as **fracdiff**.

```
library(fracdiff)
x ¡- fracdiff.sim( 100, ma=-.4, d=.3)series
fit ¡- arfima(x)
tsdisplay(residuals(fit))
```

6.8   Discussion

From the simulation results and the empirical evidence provided in the preceding sections, it is evident that the model bias and the standard errors are extremely small illustrating the viability of adopting state space modeling in estimating ARFIMA parameters.

6.9   Chapter 6 Questions

1. Do ARFIMA models depict standard long memory or generalized long memory?

2. State the requisite criteria required in ARFIMA models to depict long memory?

3. What are the two main constituent equations of a state space model framework?

4. Provide the system matrices of an ARFIMA state space configuration.

5. What requirements should an ARFIMA time series process possess in order to cast it in state space?

6. Name the two approximations that could be employed to estimate parameters of a long memory ARFIMA model using state space methodology?

7. Is it feasible and viable to use the quasi-maximum likelihood presented in this chapter to provide estimates and predictions for ARFIMA parameters of a long-range-dependent series?

8. What is the KF? Explain?

9. What role does the KF play in estimation and prediction of parameters of a long memory ARFIMA series?

10. What additional issues would you confront if the state space methodology introduced in this chapter is extended to estimate and predict parameters of a much more generalized long memory GARMA model?

7

State Space Modeling of GARMA Processes with Generalized Long Memory

Synopsis: Studies on long memory processes have become increasingly popular in the recent past in financial econometrics and mathematical statistics. Path-breaking research by Granger and Joyeux (1980) and Hosking (1981) describes a fractionally integrated autoregressive moving-average (ARFIMA) process extending the traditional autoregressive integrated moving-average (ARIMA) process. This class depicts hyperbolically decaying auto-correlations paving the way for the ARFIMA model to become a popular parametric model for long-memory time series. Thereafter, certain unique characteristics of long memory models were introduced by Andel (1986). The work of Beran (1994), Robinson (2003), and Palma (2007) provides a detailed account of long memory models in terms of statistical and econometric analysis.

7.1 Introduction

Leipus and Viano (2000) present a class of generalized fractional filters and associated theoretical aspects in modeling long memory time series with finite or infinite variance. At present, a variety of applications in macroeconomics and finance depict the relevance of fractional integration and long memory. A list of such examples includes Diebold et al. (1991) for exchange rates, Andersen et al. (2001a, b) for financial volatility series, and Baillie et al. (1996) for inflation data.

Due to the success of the ARFIMA model in practice, Gray et al. (1989) suggested another class of generalized fractional time series known as Gegenbauer ARMA (GARMA) of order (p, d, q) given by

$$\phi(B)(1 - 2uB + B^2)^d X_t = \theta(B)\epsilon_t, \tag{7.1}$$

where B is the backshift operator; u and d are real numbers; ϵ_t is white noise with zero mean and variance σ^2; and $\phi(B) = 1 - \phi_1 B - \cdots - \phi_p B^p$ and $\theta(B) = 1 + \theta_1 B + \cdots + \theta_q B^q$ are stationary AR(p) and invertible MA(q) operators.

The power spectrum of the process is given by

$$f_X(\omega) = C[4(\cos\omega - u)^2]^{-d}, \quad -\pi < \omega < \pi, \tag{7.2}$$

DOI: 10.1201/9781032627007-7

where C is a suitable constant. It is known that the process is stationary long memory when $|u| < 1$ and $0 < d < 1/2$ or $|u| = 1$ and $d < 1/4$. The long memory feature of the GARMA process was illustrated through the unbounded spectral density around the Gegenbauer frequency $\omega = \cos^{-1}(u)$, when $0 < d < 1/2$. See Chung (1996) for additional information.

As a novel contribution to the existing knowledge in terms of long memory models, truncated state space representations and the Kalman filter are considered to estimate the parameters of a GARMA time series. Further investigation of the equivalent AR and MA approximations by extending the state space modeling techniques introduced by Chan and Palma (1998) enabled proposing a procedure to find an optimal lag of the truncation through a comparative assessment. As a special case of interest and for simplicity, we consider the following stationary long memory GARMA $(0, d, 0)$ model given by

$$(1 - 2uB + B^2)^d X_t = \epsilon_t, \tag{7.3}$$

where $|u| < 1$ and $0 < d < 1/2$.

7.2 MA and AR Approximations

The Wold representation of the Gaussian Gegenbauer or GARMA $(0, d, 0)$ process given in (7.3) with $\epsilon_t \sim NID(0, \sigma^2)$ is

$$X_t = \psi(B)\epsilon_t = \sum_{j=0}^{\infty} \psi_j \epsilon_{t-j}, \tag{7.4}$$

where $\psi(B) = (1 - 2uB + B^2)^{-d}$, $\psi_0 = 1$ and the coefficients ψ_j are functionally dependent on d and u.

The Gegenbauer coefficients ψ_j have the explicit representation (Erdelyi et al., 1953, 10.9)

$$\psi_j = \sum_{q=0}^{\lfloor j/2 \rfloor} \frac{(-1)^q (2u)^{j-2q} \Gamma(d - q + j)}{q!(j - 2q)! \Gamma(d)},$$

where $\Gamma(\cdot)$ is the Gamma function and can be computed in practice using the recursive formula:

$$\psi_j = 2u \left(\frac{d-1}{j+1} \right) \psi_{j-1} - \left(\frac{2}{d-1} \right) \psi_{j-2},$$

with initial values $\psi_0 = 1$ and $\psi_1 = 2du$ (see Gould, 1974 for details).

The mth-order MA approximation of the process in (4) arises from truncating the Wold representation at lag m, i.e.,

$$X_{t,m} = \sum_{j=0}^{m} \psi_j \epsilon_{t-j}. \tag{7.5}$$

$X_{t,m}$ will be referred to as a truncated Gegenbauer process.

A state space representation of the above MA(m) model in (7.5) is

$$X_{t,m} = Z\alpha_t + \epsilon_t,$$

and

$$\alpha_{t+1} = T\alpha_t + H\epsilon_t, \tag{7.6}$$

where α_{t+1} is the $m \times 1$ state vector with elements $\alpha_{j,t+1} = E(X_{t+j,m}|\mathcal{F}_{t,m})$, $\mathcal{F}_{t,m} = \{X_{t,m}, X_{t-1,m}, \ldots\}$, and the system matrices are

$$Z = [1, 0, \ldots, 0],$$

$$T = \begin{bmatrix} 0 & 1 & 0 & \cdots & 0 \\ 0 & 0 & 1 & \cdots & 0 \\ \vdots & \vdots & \vdots & \ddots & \vdots \\ 0 & 0 & 0 & \cdots & 1 \\ 0 & 0 & 0 & \cdots & 0 \end{bmatrix},$$

$$H = \begin{bmatrix} \psi_1 \\ \psi_2 \\ \vdots \\ \psi_m \end{bmatrix},$$

with matrices Z, T, and H having dimensions $1 \times m$, $m \times m$, and $m \times 1$. See, for example, Chan and Palma (1998) for details. The specification is completed by the initial state vector distribution, $\alpha_1 \sim N(a_1, P_1)$, where $a_1 = 0$ and P_1 is the Toeplitz matrix with elements $P_{hk} = \sum_j \psi_j \psi_{j+|h-k|}$. As an alternative, the corresponding AR(m) approximation could be derived by truncating the AR(∞) representation $\pi(B)X_t = \epsilon_t$, $\pi(B) = (1 - 2uB + B^2)^d$. See Chan and Palma (1998) and Grassi and Santucci de Magistris (2014) for a comparison of the two approximations in the fractionally integrated case.

The general GARMA(p, q) case can be treated similarly: a finite-order (m) autoregressive/moving average of the polynomial is computed and cast in a suitable state space. This leads to AR(m) and MA(m) approximations that can be estimated using the Kalman filter (see Harvey, 1989, and Harvey and Proietti, 2005, for information) as presented in the next section.

7.3   Kalman Filter and QMLE

Given a sample time series $\{x_t, t = 1, \ldots, n\}$, the likelihood function of the approximating MA(m) model given in (7.6) is evaluated using the Kalman filter (KF) (see Harvey, 1989 and Durbin and Koopman, 2001), using the following set of recursions:

$$\nu_t = x_t - Za_t,$$

$$f_t = Z P_t Z',$$

$$K_t = (T P_t Z')/f_t, \tag{7.7}$$

$$a_{t+1} = T a_t + K_t \nu_t, \quad P_{t+1} = T P_t T' + H H' - K_t K_t'/f_t,$$

with $t = 1, \ldots, n$, and P_1 is as given before.

The KF returns pseudo-innovations ν_t, such that if the MA(m) approximation were the true model, $\nu_t \sim NID(0, \sigma^2 f_t)$ so that the log-likelihood of (d, u, σ^2) is (apart from a constant term)

$$\ell(d, u, \sigma^2) = -\frac{1}{2} \left(n \ln \sigma^2 + \sum_{t=1}^{n} \ln f_t + \frac{1}{\sigma^2} \sum_{t=1}^{n} \frac{\nu_t^2}{f_t} \right). \tag{7.8}$$

The scale parameter σ^2 can be concentrated out of the likelihood function, so that

$$\hat{\sigma}^2 = \frac{\sum_t \nu_t^2}{f_t},$$

and the profile likelihood is

$$\ell_{\sigma^2}(d, u) = -\frac{1}{2} \left[n(\ln \hat{\sigma}^2 + 1) + \sum_{t=1}^{n} \ln f_t \right]. \tag{7.9}$$

The maximization of (7.9) can be performed by a quasi-Newton algorithm, after a reparameterization which constrains d and u in the subset of $\mathbb{R}^2$ $[0, 0.5) \times [0, 1)$. For convenience, we use the following reparameterization: $\theta_1 = \exp(2d)/(1 + \exp(2d))$ and $\theta_2 = \exp(u)/(1 + \exp(u))$. Asymptotic standard errors for the QMLE of the parameters d and u can be obtained from the numerical second derivatives evaluated with respect to the transformed parameters θ_1 and θ_2, by using the delta method. For the properties of the QMLE, see Theorem 7.1 given at the end of Section 7.5.

7.4 An Illustrative Example

The frequency $cos^{-}1(\hat{u})$ and are side lobes due to the truncation of the MA filter (Gibbs' phenomenon). The autoregressive estimates do not suffer from the Gibbs phenomenon. This is illustrated in Figure 7.4.

Properties highlighted in the illustrative example were assessed in a large-scale simulation study. It established an optimal lag order and assessed the viability of the approximations in terms of parameter estimation and prediction. Results are presented next in Section 7.5.

7.5   Monte Carlo Experiments

Apart from the estimation of model parameters, the KF provides a mechanism to compute one step ahead and multi step ahead forecasts by generating the state predictor matrix. The matrix of one step ahead prediction forecasts is generated through the product of the state space matrix Z and the state predictor matrix. Similarly, the multistep ahead forecasts are found through the product of certain numeric quantities from the state space matrix T and the state predictor matrix through a recursive mechanism. The dual features of estimation and forecasting of the KF were utilized in a series of Monte Carlo experiments to validate the optimal estimation truncation point with the optimal truncation point for forecasting. To further assess the viability of the two approximation techniques, an iterative experiment of both techniques with minimal Monte Carlo error was conducted on a simulated GARMA$(0, d, 0)$ series with lengths (n) of 100, 200, 500, 1000, and 2000 with parameters d = 0.1, 0.2, 0.3, 0.4, 0.45, for $u = 0.8$. The experiments with 1000 iterations were executed using the 32-bit version of the MATLAB-R 2011b package on 64-bit machines that were parallelized on nine different servers with capacities ranging from 16 to 48 GB. As a result of the iterative exercise, all 10 experiments resulted in providing a unique optimal value of m in terms of the total mean square error of estimators and the in-sample forecast mean square error. In each experiment, it was intended to find the total sum of all estimator mean square errors (E-MSE) of the trace estimates of the estimator mean square error matrix and validate it by the forecasting mean square error (F-MSE) of the one-step ahead forecast. Tables 7.1–7.10 provide the results of the experiment. In terms of notation, the Monte Carlo results tables denote standard errors of estimates as SD($\bullet$), and mean square errors as MSE($\bullet$), where $\bullet$ depicts an estimator. The tables report the average of the estimated d, u, and σ, denoted $\hat{d}$, $\hat{u}$, and $\hat{\sigma}$, computed across the 1000 simulations, along with the standard error, e.g.,

$$\text{SD}(\hat{d}) = \sqrt{\frac{\sum_{r=1}^{R}(\hat{d}_r - \hat{d})^2}{R}},$$

where $\hat{d}_r$ is the QMLE of d for the rth replication and R denotes the number of replications, as well as the estimation mean square error, e.g.,

$$MSE(\hat{d}) = \Sigma_{r=1}^{R}(\hat{d}_r - d)^2/R.$$

Note: Rarely, the optimal lag order was between five lag differences (e.g., 29 between 25 and 30) in MA approximation (see Tables 7.1–7.5) and two lag differences (e.g., 10 between 9 and 11) in AR approximation (see Tables 7.6–7.10). Exact optimal approximation lag orders are in Tables 7.11 and 7.12.

TABLE 7.1

Sampling properties of the QMLE estimates of the parameters θ and σ of a Gaussian Geophase process using the MA approximations. The true generating process is $(1 - 2uB + B^2)^d X_t = \epsilon_t, \epsilon_t \sim IID\ N(0, \sigma^2)$, with $d = 0.1, u = 0.8, \sigma = 1$. The results are based on 1000 Monte Carlo replications.

n	θ			σ			MA(1)		
	$\hat{\theta}$	SD($\hat{\theta}$)	MSE($\hat{\theta}$)	$\hat{\sigma}$	SD($\hat{\sigma}$)	MSE($\hat{\sigma}$)	$\hat{\theta}$	SD($\hat{\theta}$)	MSE($\hat{\theta}$)
$n = 100$									
20	0.132	0.064	0.005	0.612	0.351	0.158	0.990	0.144	0.021
25	0.132	0.064	0.005	0.605	0.359	0.166	0.989	0.143	0.020
30	0.132	0.064	0.005	0.582	0.389	0.199	0.989	0.144	0.020
35	0.133	0.064	0.005	0.588	0.367	0.179	0.990	0.144	0.021
40	0.132	0.064	0.005	0.575	0.382	0.196	0.990	0.144	0.021
$n = 200$									
20	0.117	0.046	0.002	0.705	0.270	0.081	0.997	0.097	0.009
25	0.118	0.046	0.002	0.691	0.280	0.090	0.997	0.097	0.009
30	0.118	0.046	0.002	0.677	0.298	0.103	0.997	0.097	0.009
35	0.119	0.046	0.002	0.673	0.277	0.093	0.997	0.097	0.009
40	0.119	0.046	0.002	0.677	0.283	0.095	0.997	0.097	0.009
$n = 500$									
20	0.105	0.029	0.0008	0.788	0.165	0.027	0.995	0.063	0.004
25	0.106	0.029	0.0009	0.777	0.179	0.032	0.995	0.063	0.004
30	0.107	0.029	0.0009	0.765	0.188	0.036	0.996	0.063	0.004
35	0.107	0.030	0.0009	0.764	0.185	0.035	0.996	0.063	0.004
40	0.107	0.031	0.0010	0.751	0.197	0.041	0.996	0.063	0.004
$n = 1000$									
20	0.101	0.019	0.0003	0.841	0.074	0.007	1.002	0.045	0.002
25	0.102	0.019	0.0003	0.831	0.084	0.008	1.002	0.045	0.002
30	0.102	0.019	0.0003	0.828	0.086	0.008	1.002	0.045	0.002
35	0.102	0.019	0.0003	0.828	0.085	0.008	1.002	0.045	0.002
40	0.103	0.019	0.0004	0.820	0.099	0.010	1.002	0.045	0.002
$n = 2000$									
20	0.100	0.013	0.0001	0.849	0.037	0.003	1.001	0.031	0.001
25	0.100	0.013	0.0001	0.846	0.043	0.004	1.001	0.032	0.001
30	0.100	0.013	0.0001	0.845	0.040	0.003	1.001	0.032	0.001
35	0.101	0.013	0.0001	0.843	0.045	0.004	1.001	0.031	0.001
40	0.101	0.013	0.0001	0.841	0.042	0.003	1.001	0.031	0.001

From the simulation results of the two approximations given in Tables 7.1–7.10 for each length of the considered time series, the smallest value of E-MSE is validated by the minimal value of F-MSE. It results in an optimal truncation point (optimal m) introduced as an original concept that is independent of the series length n. Another interesting result that stems from the MA approximation tabulations above is the fact that asymptotic variance of the long memory parameter estimate $\hat{d}$ of the Gegenbauer series under consideration is approximately equal to $\frac{\pi^2}{24n}$, independent of n relating to Theorem 1 due to Chan and Palma (1998). It could be assessed by calculating

TABLE 7.2

Sampling properties of the QMLE estimates of the parameters d and u of a Gaussian Gegenbauer process using the MA approximation. The true generating process is $(1 - 2uB + B^2)^d X_t = \epsilon_t$, $\epsilon_t \sim IID\mathcal{N}(0, \sigma^2)$, with $d = 0.2$, $u = 0.8$, $\sigma = 1$. The results are based on 1000 Monte Carlo replications.

n	d			u			σ		
	$\hat{d}$	SD($\hat{d}$)	MSE($\hat{d}$)	$\hat{u}$	SD($\hat{u}$)	MSE($\hat{u}$)	$\hat{\sigma}$	SD($\hat{\sigma}$)	MSE($\hat{\sigma}$)
$n = 100$									
20	0.207	0.058	0.003	0.831	0.149	0.023	0.991	0.146	0.021
25	0.207	0.057	0.003	0.826	0.160	0.026	0.991	0.145	0.021
30	0.206	0.057	0.003	0.829	0.172	0.030	0.991	0.145	0.021
35	0.205	0.055	0.003	0.834	0.148	0.023	0.991	0.145	0.021
40	0.205	0.055	0.003	0.833	0.154	0.024	0.991	0.145	0.021
$n = 200$									
20	0.202	0.042	0.001	0.854	0.087	0.010	0.998	0.101	0.010
25	0.202	0.042	0.001	0.852	0.090	0.011	0.997	0.100	0.010
30	0.201	0.042	0.001	0.856	0.097	0.012	0.998	0.101	0.010
35	0.201	0.041	0.001	0.853	0.096	0.012	0.998	0.101	0.010
40	0.201	0.041	0.001	0.855	0.098	0.012	0.998	0.100	0.010
$n = 500$									
20	0.202	0.028	0.0007	0.859	0.038	0.005	1.005	0.065	0.004
25	0.202	0.028	0.0008	0.858	0.039	0.005	1.004	0.065	0.004
30	0.202	0.028	0.0007	0.858	0.041	0.005	1.004	0.065	0.004
35	0.202	0.028	0.0008	0.855	0.043	0.005	1.004	0.064	0.004
40	0.202	0.028	0.0007	0.859	0.049	0.006	1.004	0.064	0.004
$n = 1000$									
20	0.203	0.019	0.0003	0.856	0.020	0.003	1.012	0.046	0.002
25	0.204	0.019	0.0003	0.852	0.019	0.003	1.011	0.046	0.002
30	0.204	0.019	0.0004	0.850	0.021	0.003	1.010	0.046	0.002
35	0.204	0.020	0.0004	0.849	0.025	0.003	1.010	0.046	0.002
40	0.204	0.019	0.0004	0.850	0.030	0.003	1.010	0.046	0.002
$n = 2000$									
20	0.203	0.014	0.0002	0.855	0.012	0.003	1.011	0.033	0.001
25	0.204	0.014	0.0002	0.851	0.011	0.002	1.010	0.032	0.001
30	0.204	0.014	0.0002	0.848	0.011	0.002	1.010	0.032	0.001
35	0.205	0.014	0.0002	0.846	0.012	0.002	1.010	0.032	0.001
40	0.205	0.014	0.0002	0.844	0.012	0.002	1.010	0.032	0.001

asymptotic variance for cases $n = 100, 200, 500, 1000, 2000$ and comparing with corresponding generated values.

Furthermore, the likelihood function is a monotonically increasing function of m, although the change of the function is tiny as m gets close to the optimal m reported in the table. The latter refers either to estimation of the long memory parameter, for which as m increases the typical bias–variance trade-off is observed, the bias decreasing and the variance increasing with m, or to the out-of-sample performance of the pseudo-true predictor arising from the MA approximation. The simulation experiment shows that the optimal m (i.e., minimizing the mean square estimation and prediction error for the

TABLE 7.3

Sampling properties of the QMLE estimates of the parameters d and u of a Gaussian Gegenbauer process using the MA approximation. The true generating process is $(1 - 2uB + B^2)^d X_t = \epsilon_t$, $\epsilon_t \sim IIDN(0, \sigma^2)$, with $d = 0.3$, $u = 0.8$, $\sigma = 1$. The results are based on 1000 Monte Carlo replications.

n	$\hat{d}$	$SD(\hat{d})$	$MSE(\hat{d})$	$\hat{u}$	$SD(\hat{u})$	$MSE(\hat{u})$	$\hat{\sigma}$	$SD(\hat{\sigma})$	$MSE(\hat{\sigma})$
$n = 100$									
20	0.306	0.063	0.004	0.876	0.068	0.010	1.064	0.174	0.034
25	0.304	0.063	0.004	0.881	0.073	0.012	1.074	0.184	0.039
30	0.304	0.062	0.004	0.879	0.073	0.011	1.073	0.185	0.039
35	0.302	0.062	0.003	0.887	0.077	0.013	1.086	0.199	0.047
40	0.302	0.062	0.003	0.887	0.080	0.014	1.092	0.211	0.053
$n = 200$									
20	0.307	0.046	0.002	0.873	0.050	0.007	1.068	0.131	0.022
25	0.307	0.046	0.002	0.873	0.054	0.008	1.072	0.138	0.024
30	0.305	0.046	0.002	0.877	0.058	0.009	1.081	0.148	0.028
35	0.303	0.046	0.002	0.886	0.066	0.011	1.098	0.159	0.035
40	0.302	0.047	0.002	0.885	0.066	0.011	1.097	0.165	0.036
$n = 500$									
20	0.308	0.029	0.0009	0.857	0.030	0.004	1.058	0.085	0.010
25	0.309	0.030	0.0010	0.861	0.029	0.004	1.057	0.077	0.009
30	0.306	0.030	0.0009	0.860	0.040	0.005	1.064	0.095	0.013
35	0.308	0.031	0.0011	0.865	0.052	0.006	1.073	0.107	0.016
40	0.305	0.032	0.0011	0.860	0.047	0.005	1.069	0.106	0.016
$n = 1000$									
20	0.312	0.022	0.0006	0.857	0.014	0.003	1.051	0.051	0.005
25	0.310	0.021	0.0005	0.853	0.011	0.003	1.049	0.051	0.005
30	0.309	0.021	0.0005	0.852	0.022	0.003	1.052	0.064	0.006
35	0.313	0.023	0.0007	0.852	0.032	0.003	1.055	0.075	0.008
40	0.307	0.024	0.0006	0.848	0.027	0.003	1.053	0.070	0.007
$n = 2000$									
20	0.312	0.016	0.0004	0.857	0.008	0.003	1.052	0.036	0.004
25	0.311	0.015	0.0003	0.852	0.008	0.002	1.050	0.038	0.004
30	0.310	0.015	0.0003	0.849	0.009	0.002	1.049	0.040	0.004
35	0.314	0.016	0.0004	0.847	0.017	0.002	1.049	0.046	0.004
40	0.305	0.019	0.0004	0.844	0.011	0.002	1.048	0.043	0.004

purpose of estimating d and prediction) is rather insensitive to the lag order of MA approximation. The estimation standard error should not depend on d in large samples. As a matter of fact, for the MA approximation, the standard error does not vary relevantly and the variation is due to the Monte Carlo simulation error. The AR estimator is much more unreliable and unstable, and this justifies the variation observed: as d increases the reliability of the AR approximation decreases. In terms of asymptotic properties of an approximate MLE relating to the GARMA$(0, d, 0)$ model considered above, Chan and Palma (1998) show results as Theorems 3.1–3.3 for a specific member of the

TABLE 7.4

Sampling properties of the QMLE estimates of the parameters d and u of a Gaussian Gegenbauer process using the MA approximation. The true generating process is $(1 - 2uB + B^2)^d X_t = \epsilon_t$, $\epsilon_t \sim IIDN(0, \sigma^2)$, with $d = 0.4$, $u = 0.8$, $\sigma = 1$. The results are based on 1000 Monte Carlo replications

n	d			u			σ		
	$\bar{d}$	SD($\hat{d}$)	MSE($\hat{d}$)	$\hat{u}$	SD($\hat{u}$)	MSE($\hat{u}$)	$\hat{\sigma}$	SD($\hat{\sigma}$)	MSE($\hat{\sigma}$)
$n = 100$									
20	0.419	0.061	0.004	0.873	0.053	0.008	1.087	0.208	0.051
25	0.417	0.063	0.004	0.884	0.064	0.011	1.094	0.211	0.053
30	0.416	0.062	0.004	0.876	0.060	0.009	1.083	0.197	0.046
35	0.415	0.063	0.004	0.880	0.065	0.010	1.085	0.195	0.045
40	0.416	0.063	0.004	0.883	0.068	0.011	1.084	0.194	0.045
$n = 200$									
20	0.421	0.049	0.002	0.867	0.041	0.006	1.089	0.149	0.030
25	0.424	0.050	0.003	0.869	0.050	0.007	1.090	0.151	0.031
30	0.422	0.051	0.003	0.866	0.049	0.006	1.082	0.139	0.026
35	0.418	0.051	0.003	0.878	0.064	0.010	1.100	0.165	0.037
40	0.422	0.052	0.003	0.875	0.062	0.009	1.091	0.150	0.030
$n = 500$									
20	0.422	0.033	0.001	0.860	0.022	0.004	1.092	0.108	0.020
25	0.425	0.034	0.001	0.857	0.029	0.004	1.086	0.100	0.017
30	0.425	0.036	0.001	0.852	0.026	0.003	1.080	0.094	0.015
35	0.419	0.035	0.001	0.867	0.056	0.007	1.102	0.137	0.029
40	0.426	0.035	0.002	0.855	0.043	0.005	1.084	0.104	0.018
$n = 1000$									
20	0.424	0.027	0.001	0.858	0.014	0.003	1.091	0.083	0.015
25	0.429	0.028	0.001	0.853	0.017	0.003	1.084	0.079	0.013
30	0.429	0.029	0.001	0.847	0.011	0.002	1.076	0.066	0.010
35	0.421	0.029	0.001	0.856	0.043	0.005	1.092	0.109	0.020
40	0.431	0.028	0.001	0.844	0.023	0.002	1.073	0.067	0.010
$n = 2000$									
20	0.426	0.022	0.001	0.857	0.007	0.003	1.092	0.059	0.012
25	0.431	0.024	0.001	0.851	0.010	0.002	1.084	0.056	0.010
30	0.432	0.023	0.001	0.846	0.006	0.002	1.078	0.052	0.008
35	0.422	0.024	0.001	0.852	0.036	0.004	1.090	0.090	0.016
40	0.433	0.023	0.001	0.841	0.012	0.001	1.072	0.050	0.007

GARMA(p, d, q) family. Due to their relevance, these results are established in Theorem 7.1.

Theorem 7.1. Let n be the length of a GARMA $(0, d, 0)$ series, m the optimal lag order, $k \in \mathbb{R}^+$, and $m = n^k$. If $\widehat{\Theta}_{KF}$ is the estimator due to the Kalman filter, then under the assumptions that $0 < d < 1/2$, and $\epsilon_t \sim N(0, \sigma^2)$, the following results hold:

(a) **Consistency:** $\widehat{\Theta}_{KF} \xrightarrow{p} \Theta_0$ as $n \to \infty$, with $k > 0$, where $\Theta_0 = (d, u, \sigma)' = (\Theta_{01}, \Theta_{02}, \Theta_{03})'$ is the true vector of parameters.

TABLE 7.5

Sampling properties of the QMLE estimates of the parameters d and u of a Gaussian Gegenbauer process using the AR approximation. The true generating process is $(1 - 2uB + B2)^d X_t = \epsilon_t, \epsilon_t \sim IIDN(0, \sigma, 2)$, with $d = 0.45, u = 0.8, \sigma = 1$. The results are based on 1000 Monte Carlo replications

n	d			u			σ		
	$\hat{d}$	SD($\hat{d}$)	MSE($\hat{d}$)	$\hat{u}$	SD($\hat{u}$)	MSE($\hat{u}$)	$\hat{\sigma}$	SD($\hat{\sigma}$)	MSE($\hat{\sigma}$)
$n = 100$									
20	0.459	0.049	0.002	0.874	0.050	0.008	1.113	0.217	0.060
25	0.458	0.051	0.002	0.882	0.061	0.010	1.116	0.217	0.060
30	0.457	0.050	0.002	0.878	0.060	0.009	1.109	0.209	0.055
35	0.457	0.051	0.002	0.875	0.061	0.009	1.104	0.210	0.055
40	0.457	0.051	0.002	0.882	0.066	0.011	1.109	0.211	0.056
$n = 200$									
20	0.463	0.039	0.001	0.870	0.039	0.006	1.125	0.186	0.050
25	0.464	0.040	0.001	0.875	0.052	0.008	1.124	0.181	0.048
30	0.463	0.040	0.001	0.867	0.047	0.006	1.109	0.156	0.036
35	0.462	0.041	0.001	0.873	0.058	0.008	1.124	0.189	0.051
40	0.463	0.040	0.001	0.875	0.061	0.009	1.123	0.180	0.047
$n = 500$									
20	0.470	0.028	0.001	0.863	0.026	0.004	1.123	0.136	0.034
25	0.472	0.028	0.001	0.865	0.040	0.005	1.120	0.132	0.032
30	0.473	0.028	0.001	0.854	0.027	0.003	1.102	0.107	0.021
35	0.470	0.030	0.001	0.864	0.051	0.006	1.125	0.159	0.041
40	0.473	0.028	0.001	0.861	0.050	0.006	1.115	0.134	0.031
$n = 1000$									
20	0.474	0.022	0.001	0.860	0.018	0.004	1.123	0.105	0.026
25	0.477	0.021	0.001	0.858	0.029	0.004	1.117	0.102	0.024
30	0.478	0.021	0.001	0.849	0.013	0.002	1.100	0.076	0.016
35	0.477	0.022	0.001	0.858	0.045	0.005	1.122	0.139	0.034
40	0.478	0.021	0.001	0.851	0.038	0.004	1.107	0.103	0.022
$n = 2000$									
20	0.479	0.018	0.001	0.858	0.009	0.003	1.127	0.074	0.021
25	0.482	0.017	0.001	0.855	0.022	0.003	1.120	0.085	0.021
30	0.483	0.017	0.001	0.847	0.008	0.002	1.106	0.062	0.015
35	0.482	0.017	0.001	0.851	0.035	0.003	1.118	0.113	0.026
40	0.484	0.017	0.001	0.843	0.021	0.002	1.101	0.070	0.015

(b) **Asymptotic normality:** $\sqrt{n}(\widehat{\Theta}_{KF} - \Theta_0) \xrightarrow{d} N(0, \Sigma(\Theta_0))$, as $n \to \infty$ with $k > 1/2$,

where $\Sigma^{-1}(\Theta_0) = (\Sigma_{ij}^{-1}(\Theta_0))$ such that

$$\Sigma_{ij}^{-1}(\Theta_0) = \frac{1}{4\pi} \int_{-\pi}^{\pi} \left[\frac{\partial \log f_X(\omega)}{\partial \theta_{0i}}\right]\left[\frac{\partial \log f_X(\omega)}{\partial \theta_{0j}}\right] d\omega$$

with $f_X(\omega)$ defined as per Eq. (2),

(c) **Efficiency:** $\widehat{\Theta}_{KF}$ is an efficient estimator for $k > 1/2$.

TABLE 7.6

Sampling properties of the QMLE estimates of the parameters d and u of a Gaussian Gegenbauer process using the AR approximation. The true generating process is $(1 - 2uB + B2)^d X_t = \epsilon_t, \epsilon_t \sim IIDN(0, \sigma, 2)$, with $d = 0.1, u = 0.8, \sigma = 1$. The results are based on 1000 Monte Carlo replications

n	$\hat{d}$	d SD($\hat{d}$)	MSE($\hat{d}$)	$\hat{u}$	u SD($\hat{u}$)	MSE($\hat{u}$)	$\hat{\sigma}$	σ SD($\hat{\sigma}$)	MSE($\hat{\sigma}$)
$n = 100$									
7	0.149	0.086	0.009	0.512	0.394	0.238	0.995	0.147	0.021
9	0.150	0.086	0.010	0.497	0.385	0.240	0.996	0.148	0.021
11	0.149	0.083	0.009	0.487	0.377	0.239	0.995	0.147	0.021
13	0.151	0.084	0.009	0.480	0.368	0.237	0.996	0.147	0.021
15	0.150	0.083	0.009	0.464	0.363	0.244	0.996	0.147	0.021
$n = 200$									
7	0.128	0.059	0.004	0.574	0.334	0.162	1.002	0.102	0.010
9	0.130	0.060	0.004	0.558	0.326	0.164	1.003	0.102	0.010
11	0.131	0.059	0.004	0.539	0.321	0.171	1.003	0.102	0.010
13	0.132	0.058	0.004	0.532	0.302	0.163	1.003	0.103	0.010
15	0.133	0.058	0.004	0.517	0.302	0.171	1.004	0.102	0.010
$n = 500$									
7	0.116	0.042	0.002	0.636	0.243	0.086	1.007	0.066	0.004
9	0.118	0.042	0.002	0.612	0.246	0.095	1.008	0.066	0.004
11	0.119	0.041	0.002	0.599	0.239	0.097	1.008	0.066	0.004
13	0.120	0.041	0.002	0.589	0.236	0.100	1.008	0.066	0.004
15	0.120	0.041	0.002	0.578	0.240	0.107	1.009	0.066	0.004
$n = 1000$									
7	0.110	0.031	0.001	0.657	0.195	0.058	1.009	0.047	0.002
9	0.111	0.032	0.001	0.647	0.205	0.065	1.009	0.047	0.002
11	0.112	0.031	0.001	0.632	0.194	0.065	1.010	0.047	0.002
13	0.113	0.031	0.001	0.624	0.192	0.067	1.010	0.047	0.002
15	0.114	0.030	0.001	0.612	0.187	0.070	1.010	0.047	0.002
$n = 2000$									
7	0.105	0.025	0.0006	0.678	0.165	0.042	1.011	0.033	0.001
9	0.106	0.025	0.0006	0.669	0.175	0.047	1.012	0.033	0.001
11	0.107	0.024	0.0006	0.655	0.165	0.048	1.012	0.033	0.001
13	0.107	0.024	0.0006	0.654	0.164	0.048	1.012	0.033	0.001
15	0.108	0.023	0.0006	0.644	0.159	0.049	1.013	0.033	0.001

Proof. A detailed proof of Theorem 1 can be found in Chan and Palma (1998) and is applicable to the truncated GARMA$(0, d, 0)$ process considered in Section 7.5. $\square$

The evidence reported in this section is the outcome of a larger experiment where the parameter u varied in the admissible range, $u \in (-1, 1)$ and the GARMA$(1, d, 0)$ process $(1 - \phi B)(1 - 2uB + B^2)^d X_t = \epsilon_t, \epsilon_t \sim$ i.i.d. $N(0, 1)$was also considered in the data generating process, with $\phi \in$

TABLE 7.7

Sampling properties of the QMLE estimates of the parameters d and u of a Gaussian Gegenbauer process using the AR approximation. The true generating process is $(1 - 2uB + B^2)^d X_t = \epsilon_t, \epsilon_t \sim IIDN(0, \sigma^2)$, with $d = 0.2, u = 0.8, \sigma = 1$. The results are based on 1000 Monte Carlo replications

n	$\hat{d}$	SD($\hat{d}$)	MSE($\hat{d}$)	$\hat{u}$	SD($\hat{u}$)	MSE($\hat{u}$)	$\hat{\sigma}$	SD($\hat{\sigma}$)	MSE($\hat{\sigma}$)
$n = 100$									
7	0.234	0.092	0.009	0.637	0.271	0.100	1.028	0.152	0.024
9	0.233	0.092	0.009	0.637	0.268	0.098	1.030	0.153	0.024
11	0.231	0.088	0.008	0.636	0.262	0.095	1.031	0.152	0.024
13	0.231	0.087	0.008	0.630	0.257	0.094	1.031	0.153	0.024
15	0.231	0.087	0.008	0.623	0.258	0.098	1.031	0.152	0.024
$n = 200$									
7	0.217	0.069	0.005	0.661	0.223	0.068	1.038	0.107	0.013
9	0.219	0.069	0.005	0.649	0.231	0.075	1.040	0.107	0.013
11	0.218	0.068	0.005	0.649	0.223	0.072	1.040	0.107	0.013
13	0.220	0.066	0.004	0.635	0.218	0.074	1.041	0.107	0.013
15	0.220	0.064	0.004	0.632	0.214	0.074	1.041	0.106	0.013
$n = 500$									
7	0.207	0.049	0.002	0.684	0.167	0.041	1.044	0.067	0.006
9	0.206	0.049	0.002	0.682	0.176	0.045	1.045	0.067	0.006
11	0.209	0.048	0.002	0.667	0.166	0.045	1.047	0.067	0.006
13	0.210	0.048	0.002	0.662	0.170	0.047	1.047	0.068	0.006
15	0.210	0.047	0.002	0.659	0.166	0.047	1.047	0.068	0.006
$n = 1000$									
7	0.201	0.036	0.001	0.690	0.130	0.028	1.047	0.048	0.004
9	0.196	0.036	0.001	0.706	0.139	0.028	1.048	0.049	0.004
11	0.201	0.037	0.001	0.685	0.137	0.031	1.050	0.049	0.004
13	0.201	0.037	0.001	0.685	0.140	0.032	1.050	0.049	0.005
15	0.203	0.036	0.001	0.676	0.134	0.033	1.051	0.049	0.005
$n = 2000$									
7	0.198	0.027	0.0007	0.693	0.100	0.021	1.048	0.033	0.003
9	0.191	0.026	0.0007	0.717	0.107	0.018	1.049	0.033	0.003
11	0.197	0.028	0.0008	0.692	0.111	0.024	1.050	0.033	0.003
13	0.196	0.029	0.0008	0.696	0.114	0.023	1.051	0.034	0.003
15	0.198	0.028	0.0008	0.686	0.110	0.025	1.051	0.034	0.003

$(-1, 1)$. In the second case, ϕ is estimated assuming knowledge of the true model, i.e., the MA approximation is formed from truncating the Wold polynomial $\psi(B) = (1 - \phi B)^{-1}(1 - 2uB + B^2)^{-d}$, at truncation lag m, with the coefficients ψ_j computed by convoluting the lag polynomials $(1 - \phi B)^{-1}$ and $(1 - 2uB + B^2)^{-d}$. The main results concerning the optimal choice of m for the estimation of the memory parameter d were rather insensitive to both the true value of u and ψ. So, for clarity and synthesis, only the results referring

TABLE 7.8

Sampling properties of the QMLE estimates of the parameters d and u of a Gaussian Gegenbauer process using the AR approximation. The true generating process is $(1 - 2uB + B^2)^d X_t = \epsilon_t, \epsilon_t \sim IIDN(0, \sigma^2)$, with $d = 0.2, u = 0.8, \sigma = 1$. The results are based on 1000 Monte Carlo replications

n	d			u			σ		
	$\hat{d}$	SD($\hat{d}$)	MSE($\hat{d}$)	$\hat{u}$	SD($\hat{u}$)	MSE($\hat{u}$)	$\hat{\sigma}$	SD($\hat{\sigma}$)	MSE($\hat{\sigma}$)
$n = 100$									
7	0.298	0.091	0.008	0.716	0.214	0.053	1.084	0.172	0.036
9	0.296	0.088	0.007	0.720	0.221	0.055	1.088	0.175	0.038
11	0.291	0.089	0.008	0.732	0.222	0.054	1.089	0.173	0.038
13	0.289	0.086	0.007	0.732	0.221	0.053	1.092	0.174	0.038
15	0.288	0.085	0.007	0.733	0.220	0.052	1.091	0.172	0.038
$n = 200$									
7	0.288	0.070	0.005	0.725	0.176	0.036	1.093	0.121	0.023
9	0.284	0.068	0.005	0.733	0.182	0.037	1.095	0.122	0.024
11	0.282	0.068	0.005	0.736	0.183	0.037	1.097	0.122	0.024
13	0.280	0.067	0.004	0.741	0.181	0.036	1.099	0.123	0.024
15	0.279	0.065	0.004	0.743	0.179	0.035	1.099	0.122	0.024
$n = 500$									
7	0.287	0.051	0.002	0.714	0.130	0.024	1.099	0.076	0.015
9	0.278	0.048	0.002	0.737	0.132	0.021	1.100	0.076	0.015
11	0.279	0.051	0.003	0.737	0.141	0.023	1.102	0.076	0.016
13	0.279	0.050	0.002	0.735	0.132	0.021	1.102	0.076	0.016
15	0.278	0.048	0.002	0.736	0.134	0.022	1.104	0.077	0.016
$n = 1000$									
7	0.285	0.039	0.001	0.710	0.103	0.018	1.101	0.055	0.013
9	0.276	0.034	0.001	0.736	0.099	0.014	1.102	0.055	0.013
11	0.281	0.040	0.002	0.721	0.114	0.019	1.105	0.055	0.014
13	0.278	0.036	0.001	0.728	0.099	0.015	1.104	0.055	0.014
15	0.280	0.036	0.001	0.724	0.103	0.016	1.105	0.055	0.014
$n = 2000$									
7	0.282	0.029	0.001	0.712	0.079	0.014	1.103	0.038	0.012
9	0.273	0.025	0.001	0.740	0.073	0.008	1.104	0.038	0.012
11	0.280	0.032	0.001	0.717	0.091	0.015	1.107	0.038	0.013
13	0.277	0.027	0.001	0.727	0.074	0.010	1.107	0.038	0.012
15	0.279	0.030	0.001	0.722	0.087	0.013	1.108	0.038	0.013

to the GARMA$(0, d, 0)$ process with u fixed at 0.8 are reported. The size of the Monte Carlo errors affecting the results was checked and it was decided that 1000 replications struck a very good balance between the experiment reliability and its computational complexity.

TABLE 7.9
Sampling properties of the QMLE estimates of the parameters d and u of a Gaussian Gegenbauer process using the AR approximation. The true generating process is $(1 - 2uB + B^2)^d X_t = \epsilon_t, \epsilon_t \sim IIDN(0, \sigma^2)$, with $d = 0.4, u = 0.8, \sigma = 1$. The results are based on 1000 Monte Carlo replications

n	d			u			σ		
	$\hat{d}$	SD($\hat{d}$)	MSE($\hat{d}$)	$\hat{u}$	SD($\hat{u}$)	MSE($\hat{u}$)	$\hat{\sigma}$	SD($\hat{\sigma}$)	MSE($\hat{\sigma}$)
$n = 100$									
7	0.363	0.091	0.009	0.762	0.192	0.038	1.164	0.181	0.059
9	0.364	0.085	0.008	0.766	0.190	0.037	1.167	0.183	0.061
11	0.361	0.084	0.008	0.772	0.189	0.036	1.172	0.185	0.064
13	0.356	0.083	0.008	0.782	0.190	0.036	1.177	0.190	0.067
15	0.351	0.082	0.009	0.792	0.188	0.035	1.178	0.189	0.067
$n = 200$									
7	0.363	0.077	0.007	0.753	0.163	0.028	1.175	0.130	0.047
9	0.358	0.070	0.006	0.769	0.156	0.025	1.176	0.129	0.047
11	0.354	0.072	0.007	0.779	0.160	0.026	1.182	0.130	0.050
13	0.347	0.069	0.007	0.794	0.162	0.026	1.188	0.134	0.053
15	0.340	0.071	0.008	0.809	0.162	0.026	1.190	0.133	0.054
$n = 500$									
7	0.365	0.058	0.004	0.740	0.120	0.018	1.178	0.081	0.038
9	0.356	0.050	0.004	0.765	0.108	0.013	1.180	0.081	0.039
11	0.355	0.056	0.005	0.768	0.123	0.016	1.184	0.082	0.040
13	0.351	0.055	0.005	0.780	0.124	0.015	1.187	0.084	0.042
15	0.348	0.058	0.006	0.788	0.132	0.017	1.191	0.085	0.043
$n = 1000$									
7	0.363	0.040	0.003	0.735	0.089	0.012	1.179	0.059	0.035
9	0.355	0.035	0.003	0.762	0.079	0.007	1.181	0.059	0.036
11	0.356	0.044	0.003	0.759	0.102	0.012	1.185	0.059	0.038
13	0.352	0.041	0.004	0.769	0.096	0.010	1.187	0.061	0.038
15	0.350	0.046	0.004	0.776	0.107	0.012	1.189	0.060	0.039
$n = 2000$									
7	0.359	0.028	0.002	0.740	0.067	0.008	1.178	0.040	0.033
9	0.357	0.025	0.002	0.755	0.056	0.005	1.182	0.040	0.034
11	0.357	0.035	0.003	0.751	0.082	0.009	1.185	0.041	0.036
13	0.358	0.028	0.002	0.753	0.066	0.006	1.185	0.041	0.036
15	0.357	0.034	0.003	0.756	0.081	0.008	1.187	0.041	0.036

7.6 Comparative Assessment of Approximations

MA and AR approximations of a truncated GARMA process utilized by employing the KF on simulated data yields QMLE estimates as well as one and multistep ahead forecasts of model parameters. Rapid convergence toward optimal likelihood value with a truncated state space employing the KF defined by the Gaussian likelihood equation and mean square errors (MSE's) of

TABLE 7.10

Sampling properties of the QMLE estimates of the parameters d and u of a Gaussian Gegenbauer process using the AR approximation. The true generating process is $(1 - 2uB + B^2)^d X_t = \epsilon_t, \epsilon_t \sim IIDN(0, \sigma^2)$, with $d = 0.45, u = 0.8, \sigma = 1$. The results are based on 1000 Monte Carlo replications

n	d			u			σ		
	$\hat{d}$	SD($\hat{d}$)	MSE($\hat{d}$)	$\hat{u}$	SD($\hat{u}$)	MSE($\hat{u}$)	$\hat{\sigma}$	SD($\hat{\sigma}$)	MSE($\hat{\sigma}$)
$n = 100$									
7	0.388	0.083	0.010	0.783	0.172	0.029	1.208	0.203	0.084
9	0.385	0.079	0.010	0.804	0.167	0.028	1.220	0.210	0.092
11	0.378	0.078	0.011	0.816	0.173	0.030	1.227	0.211	0.096
13	0.374	0.079	0.012	0.824	0.172	0.030	1.233	0.213	0.099
15	0.369	0.078	0.012	0.833	0.170	0.030	1.236	0.209	0.099
$n = 200$									
7	0.388	0.072	0.008	0.778	0.145	0.021	1.216	0.142	0.067
9	0.385	0.067	0.008	0.797	0.134	0.018	1.222	0.148	0.071
11	0.377	0.069	0.010	0.814	0.143	0.020	1.232	0.150	0.076
13	0.370	0.068	0.011	0.831	0.144	0.021	1.241	0.157	0.082
15	0.365	0.068	0.011	0.841	0.144	0.022	1.244	0.155	0.083
$n = 500$									
7	0.394	0.053	0.005	0.761	0.107	0.013	1.221	0.089	0.057
9	0.391	0.049	0.005	0.779	0.099	0.010	1.224	0.089	0.058
11	0.386	0.056	0.007	0.791	0.120	0.014	1.232	0.092	0.062
13	0.381	0.056	0.007	0.804	0.121	0.014	1.240	0.098	0.067
15	0.376	0.058	0.008	0.816	0.124	0.015	1.244	0.098	0.069
$n = 1000$									
7	0.394	0.039	0.004	0.757	0.084	0.008	1.221	0.065	0.053
9	0.394	0.036	0.004	0.767	0.072	0.006	1.225	0.064	0.055
11	0.390	0.045	0.005	0.775	0.099	0.010	1.232	0.066	0.058
13	0.388	0.045	0.005	0.783	0.100	0.010	1.236	0.072	0.061
15	0.384	0.049	0.006	0.792	0.108	0.011	1.240	0.072	0.063
$n = 2000$									
7	0.390	0.027	0.004	0.761	0.062	0.005	1.222	0.044	0.051
9	0.396	0.028	0.003	0.761	0.055	0.004	1.227	0.044	0.053
11	0.394	0.035	0.004	0.763	0.077	0.007	1.230	0.045	0.055
13	0.395	0.035	0.004	0.766	0.075	0.006	1.233	0.048	0.056
15	0.393	0.037	0.004	0.768	0.081	0.007	1.234	0.048	0.057

estimators will provide a benchmark assessment to choose a better approximation option. In assessing the two approximations, lag truncation values were set at $m = [5, 10, 15, 20, 25, 30, 35, 40, 45]$ for MA and $m = [4, 7, 9, 11, 13, 15, 17]$ for AR approximations. The authors noted that the AR approximation did converge faster toward the optimal likelihood value with an approximate difference of 20 lags. However, the MA approximation provided much smaller MSE values for the estimators. Figure 7.5 provides a visual illustration for the case $d = 0.4$, $u = 0.8$, with replications $= 1000$. Furthermore, due to the easy implementation of KF recursions and simplicity of the analysis of theoretical

TABLE 7.11
Optimal values of m with $u = 0.8$ using MA approximation and 1000 replications

n	$d = 0.1$	$d = 0.2$	$d = 0.3$	$d = 0.4$	$d = 0.45$
100	35	35	30	29	35
200	35	35	33	30	30
500	35	30	35	30	30
1000	35	30	25	30	30
2000	30	30	30	30	30

Note: MA approximation—optimal lag order interval: [29, 35].

TABLE 7.12
Optimal values of m with $u = 0.8$ using AR approximation and 1000 replications

n	$d = 0.1$	$d = 0.2$	$d = 0.3$	$d = 0.4$	$d = 0.45$
100	13	13	12	9	9
200	13	11	9	9	9
500	13	10	9	9	9
1000	11	10	9	13	9
2000	12	13	9	13	9

Note: AR approximation—optimal lag order interval: [9, 13].

properties of the MLEs (as mentioned in Chan and Palma (2006) and Palma (2007) for long memory time series), the MA representation became the approximation of choice over its competing rival. Another reason for choosing the MA approximation was the advent of a smaller error variance than for the AR approximation in terms of a differenced long memory Gegenbauer time series. By summarizing the results in Tables 7.1–7.10 for both the approximations, the optimal value of the truncation point for each length of the same series is shown in Tables 7.11 and 7.12.

Remark. It is evident that the approximations presented in this section are appropriate for a Wold type linear series driven by Gaussian white noise that could be fitted with a long memory Gegenbauer process depicting hyperbolically decaying autocorrelation/partial autocorrelation functions, and an unbounded spectral density away from the origin. The implementation efficiency in terms of processing time is achieved up to approximately $m = 70$ lags.

Note: Taking into consideration the overall attributes of the two approximations from Sections 7.5 and 7.6, the better-performing MA approximation became the chosen option. Therefore, the majority of the results presented below are based on the MA approximation. Further results based on the AR approximation will be clearly labeled adjacent to the presented information. As confirmatory examples of both the approximations, results of the case with $d = 0.4$, $u = 0.8$, and Replications $= 1000$ for varying lengths of n are provided in Tables 7.13 and 7.14. By investigating the results of Tables 7.11 and 7.12 for the MA approximation, it is clear that the optimal truncation point for the long memory Gegenbauer series under consideration will lie in the interval [29, 35] for any value of n. In addition to the experimental results given above for comparison purposes, several other Monte Carlo experiments with MA approximation were conducted to estimate the long memory and Gegenbauer frequency parameters of the desired GARMA model. This was executed using combinations of the sets of values $d = [0.1, 0.2, 0.3, 0.4, 0.45]$, $m = [6, 14]$, $n = [1000, 5000, 10\,000]$, and Replications $= [100, 200, 500, 1000]$. In one such experiment, the estimated value and the standard error for the long memory model parameter $d = 0.45$ of a series with $n = 1000$ after 100 replications were far superior to a similar result from a two factor GARMA model with the same length and 1000 replications presented in Bisaglia et al. (2003). Furthermore, estimates comparable with the results given by Chan and Palma (1998) for the ARFIMA model were obtained through additional Monte Carlo experiments. All the simulation estimates were comparable with similar GARMA model parameter estimates shown in Gray et al. (1989), Chung (1996), and Beaumont

TABLE 7.13

MA approximation $d = 0.4$, $u = 0.8$ replications 1000

m	$n = 100/$ $F - MSE$	$n = 200/$ $F - MSE$	$n = 500/$ $F - MSE$	$n = 1000/$ $F - MSE$	$n = 2000/$ $F - MSE$
20	1.2487	1.2062	1.0424	1.0479	1.1374
25	1.2497	1.1913	1.0421	1.0453	1.1319
30	1.2482	1.1909	1.0370	1.0416	1.1274
35	1.2395	1.1952	1.0395	1.0453	1.1293
40	1.2505	1.1868	1.0383	1.0443	1.1234

TABLE 7.14

MA approximation $d = 0.4$, $u = 0.8$ replications 1000

m	$n = 100/$ $F - MSE$	$n = 200/$ $F - MSE$	$n = 500/$ $F - MSE$	$n = 1000/$ $F - MSE$	$n = 2000/$ $F - MSE$
7	1.1726	1.1438	1.1582	1.1769	1.1810
9	1.1435	1.0997	1.1529	1.1764	1.1869
11	1.1801	1.1372	1.1547	1.1751	1.1878
13	1.1792	1.1207	1.1515	1.1736	1.1854
15	1.1912	1.0999	1.1548	1.1743	1.1872

and Ramachandran (2001). Since the truncated state space approach with the MA approximation has been identified and assessed as a feasible procedure creating robust estimators validated through an optimal lag truncation value, it was incorporated into two segments of a real data application consisting of sunspots illustrated by Wolfer and Tong. The sunspots were chosen as the desired empirical data source, since they had been utilized for applications of various GARMA models in the past based on the work of Gray et al. (1989), Chung (1996), and Beaumont and Ramachandran (2001). The analytical results of the empirical applications are the focal point of the next section.

7.7   Results of Empirical Applications

The truncated state space approach incorporating the MA approximation after being classified as a viable and feasible technique in estimating and forecasting with respect to Gegenbauer long memory time series models was tested with a real application from the popular versions of sunspot time series. In this section, for comparative purposes in certain instances the AR approximation results are included as well. The mean square error (MSE) and the mean absolute error (MAE) were used as benchmark measures in the comparison tables given below, where

$$MSE = (1/N)\Sigma_{i=1}^{N}e_i^2, \tag{7.10}$$

$$MSE = (1/N)\Sigma_{i=1}^{N}|e_i^2|, \tag{7.11}$$

with e_i defined as the forecast error, and N the length of the series of forecast errors.

A long memory Gegenbauer model is fitted to Wolfer's series with length 176 (source: Waldmeier, 1961) and Tong's series with length 289 (source: Tong, 1990) comprising of sunspots data. The results are given in Tables 7.15–7.17. Reference plots of the two sunspots series are shown in Figure 7.6.

In order to make a forecast assessment of Wolfer's sunspot data series, a rolling window of 100 data points was utilized with eight-step ahead forecast predictions at each iteration to achieve predicted values for the entire series. It yielded mean square forecast errors that were better up to the five step ahead forecast than the values given in Morris (1977) for the same series using a similar forecasting technique. In Table 7.16, the comparative results of the MSE values beginning at the third step ahead forecast beyond the minimum year of a sunspot cycle is recorded, since Morris (1977) begins the assessment at the same evaluation point. To further illustrate the forecasting efficiency of the method proposed in this paper, a comparative evaluation was made with respect to the results provided in Zhang (2003) and Bijari and Khashei (2011) with respect to Wolfer's sunspots by utilizing the same data and are shown below in Table 7.17.

TABLE 7.15

Estimation results for Wolfer's/Tong's sunspot series (standard errors are given within brackets)

Series	$\hat{d}$(MA)	$\hat{u}$(MA)	$\hat{d}$(AR)	$\hat{u}$(AR)
Wolfer's sunspots	0.49 (0.002)	0.85 (0.003)	0.49 (0.002)	0.89 (0.006)
Tong's sunspots	0.49 (0.002)	0.84 (0.002)	0.49 (0.002)	0.89 (0.005)

Note: The estimated values $\hat{d}$ for both the sunspot series were less than 0.5 with $\hat{u} < 1$ up to four decimal places illustrating stationarity. Under the AR approximation, both sunspot series correspond to periods of around 13.2 observations. However, the MA approximation delivers periods of 11.3 and 10.9 observations, respectively, for Wolfer's and Tong's sunspot series. Therefore, Tong's series with a larger length provides a better estimate under the MA approximation that is closer to the approximate theoretical sunspot cycle periodicity of 10 observations. Yet again, the MA approximation outperforms its AR rival through empirical evidence.

TABLE 7.16

Comparative assessment of forecast errors for Wolfer's sunspot series with Morris's model

Year after minimum	3	4	5
MSE for Morris's AR model	580.5	537.9	345.5
MSE for proposed model	117.1	174.1	268.6

TABLE 7.17

Comparative assessment of forecast errors for Wolfer's sunspot series with optimal autoregressive integrated moving average (ARIMA) models

Model	67 time points ahead—MAE
AR(9)-ARIMA	13.033
Zhang's hybrid ARIMA	12.780
Bijari–Khashei hybrid ARIMA	11.446
Proposed model (MA approximation) at $m = 30$	12.877

From the results of Table 7.16, it is clear that the forecast model presented in this paper performs better than the AR model introduced in Morris (1977) in terms of the forecast mean square errors ranging from three- to five-step ahead predictions. In his paper, it is acknowledged that the utilized traditional AR model up to 30 lags does not yield satisfactory forecast MSE's. He compensates for it by mixing it with an outburst regression model through

a weighting mechanism. Also, he does not mention an optimal lag truncation value. In the proposed model as the lags increase until an optimal order
of $m = 30$ the typical bias–variance trade-off is observed. Thereby, it minimizes the prediction error to a great extent resulting in MSEs significantly
smaller than corresponding MSEs of Morris's AR model. Furthermore, from
the comparison results of Table 7.17, it is apparent that the proposed model
performs better than the traditional AR(9)-ARIMA time series model given
in the literature of Zhang (2003) and Bijari and Khashei (2011) with a lag
order of $m = 30$ further corroborating the optimal lag interval of $[29, 35]$ established by the preceding simulations. Furthermore, in a forecasting sense
the Gegenbauer model performs as efficiently as the Zhang (2003) and Bijari and Khashei (2011) hybrid models without the propelling of any neural
networks. It provides the implication that a hybrid Gegenbauer long memory
model with neural networks may outperform the recently established modern
hybrid models. Therefore, it will clearly match or outperform all the three
ARIMA models given in Table 7.17 with respect to the 67 time points ahead
prediction within the proposed optimal lag truncation range of $[29, 35]$ at $m =
30$ illustrating and corroborating the long memory property. The investigative
results arrived at in the preceding sections are summarized and become the
conclusion in the final section.

Note: Incorporating the seasonal operator to the state space configuration
of the GARMA$(0, d, 0)$ model introduced above would result in a Gegenbauer autoregressive seasonal moving average time series model denoted as
a GARSMA$(0, d, 0)(0, D', 0)$ process. Theoretical, simulated and empirical
evidence linked with the process is presented below as an extension of the
GARMA process.

7.8 Seasonal Operator

Let s' be the seasonal period. Then, the seasonal operator introduced in Chapter 2 is considered for the corresponding theoretical developments. Details of
the derived factorizations are provided next in the discussion.

7.8.1 Factorization of the seasonal operator

Consider the seasonal lag polynomial $(1 - B^{s'})$, with B the backshift operator.
It is known that the s'-th roots of unity are given by

$$\zeta_k = cos(\omega_k) + isin(\omega_k), \quad k = 0, 1, 2, ..., s' - 1, \tag{7.12}$$

where $\omega_k = 2k\pi/s'$.

This implies that we can write:

$$(1 - B^{s'}) = \varrho_{s'} \prod_{k=0}^{s'-1} (\zeta_k - B), \tag{7.13}$$

for some constant $\varrho_{s'}$.

Comparing the coefficients of $B^{s'}$, it is clear that

$$\varrho_{s'} = \begin{cases} 1, & \text{if } s' \text{ is odd} \\ -1, & \text{if } s' \text{ is even.} \end{cases}$$

For every $k > 0$, we have $\zeta_{s'-k} = cos(2\pi - \omega_k) + isin(2\pi - \omega_k) = cos(\omega_k - isin(\omega_k))$, and that $(\zeta_k - B)(\zeta_{s'-k} - B) = cos^2(\omega_k) + sin^2(\omega_k) - 2cos(\omega_k)B + B^2 = 1 - 2cos(\omega_k)B + B^2$.

Case 1: s' is odd

If s' is odd, pairing off is done for all factors except when $k = 0$. It is clear that, $1 - B^{s'} = \prod_{k=0}^{s'-1}(\zeta_k - B) = (\zeta_0 - B)\prod_{k=1}^{(s'-1)/2}[(\zeta_k - B)(\zeta_{s'-k} - B)]$
$= (1 - B)\prod_{k=1}^{(s'-1)/2}(1 - 2cos(\omega_k)B + B^2)$.

It is easy to verify that $1 - B = (1 - 2B + B^2)^{1/2} = (1 - 2cos(0)B + B^2)^{1/2} = (1 - 2cos(\omega_0)B + B^2)^{1/2}$, and hence

$$1 - B^{s'} = (1 - 2cos(\omega_0)B + B^2)^{1/2} \prod_{k=1}^{(s'-1)/2} (1 - 2cos(\omega_k)B + B^2). \tag{7.14}$$

Case 2: s' is even

If s' is even, pairing off is done for all factors except $k = 0$ and $k = s'/2$. Noting that $\zeta_0 = 1$ and $\zeta_{s'/2} = -1$, it is clear that,
$1 - B^{s'} = \prod_{k=0}^{s'-1}(\zeta_k - B) = -(\zeta_0 - B)(\zeta_{s'/2} - B)\prod_{k=1}^{(s'/2)-1}[(\zeta_k - B)(\zeta_{s'-k} - B)]$
$= -(1 - B)(-1 - B)\prod_{k=1}^{(s'/2)-1}(1 - 2cos(\omega_k)B + B^2)$
$= (1 - B)(1 + B)\prod_{k=1}^{(s'/2)-1}(1 - 2cos(\omega_k)B + B^2)$.

As before
$1 + B = (1 + 2B + B^2)^{1/2} = (1 - 2cos(\pi)B + B^2)^{1/2} = (1 - 2cos(\omega_{s'/2})B + B^2)^{1/2}$,
and hence

$$1 - B^{s'} = (1 - 2cos(\omega_0)B + B^2)^{1/2}(1 - 2cos(\omega_{s'/2})B + B^2)^{1/2}$$
$$\times \prod_{k=1}^{(s'/2)-1} (1 - 2cos(\omega_k)B + B^2). \tag{7.15}$$

The next section considers the use of equations defined above in seasonal modeling.

7.8.2 The GARSMA(0,d,0)x(0,$D_{s'}$,0) model

Suppose that $\{Y_t\}$ is a seasonal time series with period s' that can be transformed into

$$X_t = (1 - B^{s'})^{D_{s'}} Y_t$$

to remove any seasonal components, where $0 < D_{s'} < \frac{1}{2}$.

Then, GARMA$(0, d, 0)$ series with seasonality can be written as

$$(1 - 2uB + B^2)^d (1 - B^{s'})^{D_{s'}} X_t = \epsilon_t, \quad 0 < d < \frac{1}{2} \quad and \quad |u| < 1. \quad (7.16)$$

Extending the notion of an ARFISMA series can be defined as a GARSMA(0,d,0)x(0,$D_{s'}$,0) model.

Let

$$(1 - 2uB + B^2)^d (1 - B^{s'})^{D_{s'}} = \sum_{j=1}^{\infty} \pi_j' B^{s'j}, \quad (7.17)$$

$$(1 - 2uB + B^2)^{-d} (1 - B^{s'})^{-D_{s'}} = \sum_{j=1}^{\infty} \psi_j' B^{s'j}, \quad (7.18)$$

where ψ_j' and π_j' are the corresponding coefficients of each

$$(1 - B^{s'})^{-D_{s'}} = \sum_{j=1}^{\infty} C_j' B^{s'j}$$

expansion. The following lemma is useful for later reference:

Lemma 7.1. A GARSMA(0,d,0)x(0,$D_{s'}$,0) process of (7.5) is stationary and long memory when $|u| < 1$, $0 < d < 1/2$ and $0 < D_{s'} < 1/2$.

7.8.3 Properties of GARSMA(0,d,0)x(0,$D_{s'}$,0) model

A Wold's representation of a MA approximation-based GARSMA$(0, d, 0) \times (0, D_{s'}, 0)$ model is

$$X_t = \psi'(B)\epsilon_t = \sum_{j=0}^{\infty} \psi_j' \epsilon_{t-j}, \quad (7.19)$$

Therefore, ψ_j' will be a convolution of coefficients C_j' and C_j such that $C_j' = \Gamma(s' \times j + d)/\Gamma(d)\Gamma(s' \times j + 1)$,
and C_j are Gegenbauer coefficients.
where $\Gamma(\bullet)$ is the Gamma function.
Refer Palma (2007) for details.
Now, we consider the state space representation of a GARSMA$(0, d, 0) \times (0, D_{s'}, 0)$ process
Extending state space modeling to assess the properties of a GARSMA $(0, d, 0) \times (0, D_{s'}, 0)$ process remains a viable exercise and will be the discussion topic of the next section.

7.9 State Space Representation of a GARSMA(0, d, 0) × (0, $D_{s'}$, 0) Process

As in the nonseasonal case, the m-th-order truncated MA approximation or the Wold representation of (7.8) at lag m is

$$X_{t,m} = \sum_{j=0}^{m} \psi'_j \epsilon_{t-j} \qquad (7.20)$$

Following Chan and Palma (1998) and Dissanayake et al (2014a), the state space representation of the MA(m) model is given by measurement/observation and transition/state equations as:

$$
\begin{aligned}
X_{t,m} &= Z\alpha_t + \epsilon_t, \\
\alpha_{t+1} &= T\alpha_t + H\epsilon_t,
\end{aligned}
\qquad (7.21)
$$

where α_{t+1} is the $m \times 1$ state vector with elements $\alpha_{j,t+1} = E(X_{t+j,m}|\mathcal{F}_t), \mathcal{F}_t = \{X_{t,m}, X_{t-1,m}, \ldots\}$.

Following Chan and Palma (1998), it can be shown that the system matrices of the constructed state space configuration are

$$
Z = [1, 0, \ldots, 0], \quad
T = \begin{bmatrix}
0 & 1 & 0 & \cdots & 0 \\
0 & 0 & 1 & \ddots & 0 \\
\vdots & \vdots & \ddots & \ddots & 0 \\
\vdots & \cdots & \cdots & 0 & 1 \\
0 & 0 & \cdots & \cdots & 0
\end{bmatrix}, \quad
H = \begin{bmatrix}
\psi'_1 \\
\psi'_2 \\
\vdots \\
\vdots \\
\psi'_m
\end{bmatrix}
$$

H is a column vector of convoluted seasonal Gegenbauer coefficients and $G = [1]$.

It is possible to derive the corresponding AR(m) approximation by truncating the AR(∞) representation $\pi'(B)X_t = \epsilon_t$, $\pi'(B) = (1 - B^{s'})^{D_{s'}}(1 - 2uB + B^2)^d$ following Chan and Palma (1998) and Grassi and De Magistris (2014).

An estimation procedure based on the state space models originally developed by Kalman (1961) and used in Chapter 3 for a GARMA$(0, d, 0)$ model is extended to the system given in (7.10) using KF recursions.

7.9.1 QML Estimation through KF Recursion

For a time series $\{x_t, t = 1, \ldots, n\}$, the approximate likelihood function of an MA(m) model evaluated using the KF is given below:

$$
\begin{aligned}
\nu_t &= x_t - Za_t, & f_t &= ZP_tZ', \\
& & K_t &= (TP_tZ')/f_t, \\
a_{t+1} &= Ta_t + K_t\nu_t, & P_{t+1} &= TP_tT' + HH' - K_tK'_t/f_t.
\end{aligned}
\qquad (7.22)
$$

The KF returns pseudo-innovations ν_t, creating a log-likelihood of $(d, u, D_{s'}, \sigma^2)$ (apart from constant term)

$$\ell(d, u, D_{s'}, \sigma^2) = -\frac{1}{2}\left(n \ln \sigma^2 + \sum_{t=1}^{n} \ln f_t + \frac{1}{\sigma^2}\sum_{t=1}^{n} \frac{\nu_t^2}{f_t}\right) \qquad (7.23)$$

and the profile likelihood of

$$\ell_{\sigma^2}(d, u, D_{s'}) = -\frac{1}{2}\left[n(\ln \hat{\sigma}^2 + 1) + \sum_{t=1}^{n} \ln f_t\right] \qquad (7.24)$$

Remark 7.1. Adopting the methodology introduced in Sections 5.2 and 5.3 of this chapter in order to assess finite sample performance of QML estimates in terms of estimation and forecasting a number of Monte Carlo simulations were performed. In that context, a GARSMA(0,d,0)x(0,$D_{s'}$,0) process having a monthly periodicity of $s' = 12$ is considered throughout this chapter to appropriately assess the model. The chosen special case GARSMA(0,d,0)x(0,$D_{s'}$,0) model is

$$(1 - 2uB + B^2)^d(1 - B^{12})^{D_{s'}} X_t = \epsilon_t. \qquad (7.25)$$

Monte Carlo evidence for small lengths ($n = 100, 200$) of the desired model was performed in lieu of it, and the results are presented in the next section.

7.10   Monte Carlo Evidence

Monte Carlo experiments were executed to assess and corroborate the developed theory of the desired GARSMA(0,d,0)x(0,$D_{s'}$,0) model. The results are provided in Tables 7.18–7.29. In order to arrive at an optimal lag order (m), the total estimator mean square error (E.MSE) was validated by a rolling forecast one step ahead prediction mean square error (F.MSE), where $E.MSE = MSE(\hat{d}) + MSE(\hat{u}) + MSE(\hat{D}_{s'}) + MSE(\hat{\sigma})$. $d = [0.1, 0.3, 0.45]$, $D_{s'} = [0.1, 0.3, 0.45]$ and $n = [100, 200]$.

Note: By carefully inspecting Tables 7.18–7.29 it could be observed that the E.MSE value decrease and subsequently increase with a turning point (first lowest dip) in terms of the lag order. It is also evident that a similar behavior takes place with respect to the F.MSE at the same lag order. Therefore the total model estimator mean square error is validated by the predictive accuracy resulting in an optimal lag order (m). The exercise is similar to what was done for a GARMA$(0, d, 0)$ model in Chapter 3 and Dissanayake et al. (2014a).

Note: In the next segment, a comparative meta-analysis in terms of optimal lag order of the above results in terms of the two approximations and corresponding results of a GARMA model are provided.

TABLE 7.18

MA approximation with $d = 0.1$, $u = 0.8$, $D_{s'} = 0.45$, and replications = 1000

			$n = 100$			
m	$\hat{d}$	SD($\hat{d}$)	MSE($\hat{d}$)	$\hat{u}$	SD($\hat{u}$)	MSE($\hat{u}$)
40	0.080	0.063	0.004	0.697	0.364	0.153
45	0.087	0.063	0.004	0.687	0.346	0.145
50	0.087	0.063	0.004	0.696	0.348	0.145

				$n = 100$				
m	$\hat{\sigma}$	SD($\hat{\sigma}$)	MSE($\hat{\sigma}$)	$\hat{D_{s'}}$	SD($\hat{D_{s'}}$)	MSE($\hat{D_{s'}}$)	E. MSE	F. MSE
40	0.995	0.149	0.022	0.443	0.016	0.003	0.182	1.0805
45	0.991	0.150	0.022	0.428	0.017	0.004	0.175	1.0804
50	0.991	0.150	0.022	0.431	0.015	0.007	0.178	1.1171

TABLE 7.19

AR approximation with $d = 0.1$, $u = 0.8$, $D_{s'} = 0.45$, and replications = 1000

			$n = 100$			
m	$\hat{d}$	SD($\hat{d}$)	MSE($\hat{d}$)	$\hat{u}$	SD($\hat{u}$)	MSE($\hat{u}$)
30	0.083	0.075	0.007	0.686	0.157	0.121
35	0.083	0.072	0.007	0.676	0.158	0.119
40	0.081	0.067	0.007	0.678	0.149	0.123

				$n = 100$				
m	$\hat{\sigma}$	SD($\hat{\sigma}$)	MSE($\hat{\sigma}$)	$\hat{D_{s'}}$	SD($\hat{D_{s'}}$)	MSE($\hat{D_{s'}}$)	E. MSE	F. MSE
30	1.020	0.151	0.023	0.386	0.074	0.009	0.160	0.991
35	1.021	0.150	0.023	0.384	0.074	0.009	0.158	0.988
40	1.023	0.152	0.023	0.385	0.072	0.009	0.162	0.992

7.10.1 Comparison of AR and MA approximations

See Tables 7.30 and 7.31.

Optimal lag order using MA approximation – $[45, 55]$

Optimal lag order using AR approximation – $[35, 45]$

Remark 7.2. Next, a comparison of lag orders with GARMA$(0, d, 0)$ model is provided.

TABLE 7.20
MA approximation with $d = 0.1$, $u = 0.8$, $D_{s'} = 0.45$, and replications = 1000

			$n = 200$				
m	$\hat{d}$	SD($\hat{d}$)	MSE($\hat{d}$)	$\hat{u}$	SD($\hat{u}$)	MSE($\hat{u}$)	
45	0.079	0.051	0.003	0.638	0.240	0.085	
50	0.079	0.048	0.002	0.669	0.245	0.072	
55	0.072	0.052	0.003	0.562	0.243	0.080	

			$n = 200$				
m	$\hat{\sigma}$	SD($\hat{\sigma}$)	MSE($\hat{\sigma}$)	$\hat{D}_{s'}$	SD($\hat{D}_{s'}$)	MSE($\hat{D}_{s'}$)	E. MSE	F. MSE
45	1.008	0.107	0.011	0.389	0.015	0.002	0.101	0.974
50	1.007	0.107	0.011	0.382	0.015	0.002	0.087	0.971
55	1.008	0.110	0.012	0.369	0.014	0.005	0.103	0.977

TABLE 7.21
AR approximation with $d = 0.1$, $u = 0.8$, $D_{s'} = 0.45$, and replications = 1000

			$n = 200$				
m	$\hat{d}$	SD($\hat{d}$)	MSE($\hat{d}$)	$\hat{u}$	SD($\hat{u}$)	MSE($\hat{u}$)	
35	0.061	0.052	0.006	0.707	0.109	0.083	
40	0.061	0.050	0.006	0.694	0.107	0.001	
45	0.060	0.048	0.006	0.696	0.105	0.036	

			$n = 200$				
m	$\hat{\sigma}$	SD($\hat{\sigma}$)	MSE($\hat{\sigma}$)	$\hat{D}_{s'}$	SD($\hat{D}_{s'}$)	MSE($\hat{D}_{s'}$)	E. MSE	F. MSE
35	1.027	0.104	0.011	0.422	0.012	0.006	0.106	1.0066
40	1.028	0.104	0.011	0.419	0.012	0.008	0.026	1.0065
45	1.028	0.105	0.011	0.418	0.012	0.008	0.061	1.0066

7.10.2 A meta-analysis with GARMA$(0, d, 0)$ model results

From the above results, it is clear that the introduction of the seasonal filter has an impact in extending the optimal lag order (Table 7.32).

Since the difference in the lag order intervals was not great and due to the fact that the MA approximation was returning smaller MSE's for the estimators, it was chosen as the better representation. Therefore, the MA approximation within the established optimal lag order interval was utilized to assess two real applications.

TABLE 7.22
MA approximation with $d = 0.3$, $u = 0.8$, $D_{s'} = 0.3$, and replications $= 1000$

				$n = 100$			
m	$\hat{d}$	SD($\hat{d}$)	MSE($\hat{d}$)	$\hat{u}$	SD($\hat{u}$)	MSE($\hat{u}$)
50	0.305	0.052	0.002	0.856	0.091	0.014
55	0.311	0.050	0.002	0.845	0.096	0.011
60	0.304	0.051	0.002	0.854	0.109	0.015

				$n = 100$				
m	$\hat{\sigma}$	SD($\hat{\sigma}$)	MSE($\hat{\sigma}$)	$\hat{D_{s'}}$	SD($\hat{D_{s'}}$)	MSE($\hat{D_{s'}}$)	E. MSE	F. MSE
50	1.14	0.133	0.037	0.307	0.024	0.003	0.056	1.0805
55	1.12	0.114	0.028	0.321	0.060	0.004	0.045	1.0804
60	1.14	0.133	0.037	0.313	0.054	0.003	0.057	1.1171

TABLE 7.23
AR approximation with $d = 0.3$, $u = 0.8$, $D_{s'} = 0.3$, and replications $= 1000$

				$n = 100$			
m	$\hat{d}$	SD($\hat{d}$)	MSE($\hat{d}$)	$\hat{u}$	SD($\hat{u}$)	MSE($\hat{u}$)
40	0.281	0.095	0.009	0.804	0.273	0.074
45	0.286	0.086	0.007	0.818	0.207	0.043
50	0.258	0.112	0.014	0.729	0.470	0.226

				$n = 100$				
m	$\hat{\sigma}$	SD($\hat{\sigma}$)	MSE($\hat{\sigma}$)	$\hat{D_{s'}}$	SD($\hat{D_{s'}}$)	MSE($\hat{D_{s'}}$)	E. MSE	F. MSE
40	1.077	0.245	0.066	0.369	0.163	0.003	0.152	0.997
45	1.078	0.244	0.066	0.380	0.157	0.003	0.119	0.993
50	1.116	0.283	0.094	0.391	0.151	0.003	0.337	0.995

7.11 Empirical applications

7.11.1 El'Nino data

Literature on Climatology records El'Nino series observations based on seasonal monthly drought outlook. In that context, a current El' Nino series was chosen as the first real application to fit the GARSMA$(0,d,0)$x$(0,D_{s'},0)$ model introduced in this chapter, since Besse et al. (2000) and others had used an El'Nino series to make forecasts about some functional climatic variations

TABLE 7.24
MA approximation with $d = 0.3$, $u = 0.8$, $D_{s'} = 0.3$, and replications = 1000

m	$\hat{d}$	SD($\hat{d}$)	MSE($\hat{d}$)	$\hat{u}$	SD($\hat{u}$)	MSE($\hat{u}$)
			$n = 200$			
40	0.267	0.029	0.002	0.969	0.069	0.003
45	0.307	0.028	0.0008	0.844	0.072	0.007
50	0.309	0.026	0.0008	0.857	0.075	0.009

m	$\hat{\sigma}$	SD($\hat{\sigma}$)	MSE($\hat{\sigma}$)	$\hat{D}_{s'}$	SD($\hat{D}_{s'}$)	MSE($\hat{D}_{s'}$)	E. MSE	F. MSE
				$n = 200$				
40	1.284	0.112	0.033	0.362	0.019	0.001	0.039	1.110
45	1.114	0.101	0.023	0.371	0.051	0.003	0.033	1.108
50	1.130	0.119	0.031	0.384	0.046	0.003	0.043	1.114

TABLE 7.25
AR approximation with $d = 0.3$, $u = 0.8$, $D_{s'} = 0.3$, and replications = 1000

m	$\hat{d}$	SD($\hat{d}$)	MSE($\hat{d}$)	$\hat{u}$	SD($\hat{u}$)	MSE($\hat{u}$)
			$n = 200$			
40	0.265	0.058	0.004	0.951	0.131	0.040
45	0.316	0.070	0.005	0.820	0.106	0.011
50	0.266	0.066	0.005	0.936	0.106	0.027

m	$\hat{\sigma}$	SD($\hat{\sigma}$)	MSE($\hat{\sigma}$)	$\hat{D}_{s'}$	SD($\hat{D}_{s'}$)	MSE($\hat{D}_{s'}$)	E. MSE	F. MSE
				$n = 200$				
40	1.240	0.215	0.042	0.323	0.117	0.001	0.087	1.146
45	1.104	0.178	0.042	0.389	0.149	0.002	0.060	0.917
50	1.219	0.220	0.090	0.392	0.150	0.002	0.124	1.150

using an autoregressive process. The following Table 7.33 provides the estimation results with the standard error estimates shown within brackets.

Note: The optimal series model will be $(1 - 2 \times 0.971B + B^2)^{0.269}(1 - B^{12})^{0.324}X_t = \epsilon_t$. The standard error estimates of both the memory parameters are small and reasonably close in value. The estimates fall within the bounds of the stipulated values of the introduced model depicting seasonality and long memory. The optimal forecasts are achieved at comparatively large lag orders raising questions about further improving and enhancing the model in terms of cost, processing speed, affordability, and efficiency.

TABLE 7.26

MA approximation with $d = 0.45$, $u = 0.8$, $D_{s'} = 0.1$, and replications $= 1000$

			$n = 100$			
m	$\hat{d}$	SD($\hat{d}$)	MSE($\hat{d}$)	$\hat{u}$	SD($\hat{u}$)	MSE($\hat{u}$)
40	0.442	0.041	0.002	0.869	0.061	0.008
45	0.457	0.040	0.002	0.818	0.058	0.007
50	0.460	0.040	0.003	0.910	0.062	0.007

				$n = 100$				
m	$\hat{\sigma}$	SD($\hat{\sigma}$)	MSE($\hat{\sigma}$)	$\hat{D_{s'}}$	SD($\hat{D_{s'}}$)	MSE($\hat{D_{s'}}$)	E. MSE	F. MSE
40	1.66	0.265	0.081	0.107	0.038	0.003	0.094	1.011
45	1.71	0.216	0.077	0.121	0.023	0.004	0.090	0.932
50	1.60	0.298	0.078	0.113	0.016	0.003	0.091	0.988

TABLE 7.27

AR approximation with $d = 0.45$, $u = 0.8$, $D_{s'} = 0.1$, and iterations $= 1000$

			$n = 100$			
m	$\hat{d}$	SD($\hat{d}$)	MSE($\hat{d}$)	$\hat{u}$	SD($\hat{u}$)	MSE($\hat{u}$)
30	0.458	0.096	0.013	0.649	0.354	0.062
35	0.455	0.099	0.013	0.634	0.348	0.060
40	0.457	0.095	0.013	0.654	0.314	0.066

				$n = 100$				
m	$\hat{\sigma}$	SD($\hat{\sigma}$)	MSE($\hat{\sigma}$)	$\hat{D_{s'}}$	SD($\hat{D_{s'}}$)	MSE($\hat{D_{s'}}$)	E. MSE	F. MSE
30	1.38	0.355	0.099	0.104	0.089	0.017	0.191	0.977
35	1.40	0.355	0.080	0.100	0.089	0.018	0.171	0.960
40	1.41	0.361	0.088	0.103	0.082	0.018	0.185	0.974

TABLE 7.28

MA approximation with $d = 0.45$, $u = 0.8$, $D_{s'} = 0.1$, and iterations $= 1000$

			$n = 200$			
m	$\hat{d}$	SD($\hat{d}$)	MSE($\hat{d}$)	$\hat{u}$	SD($\hat{u}$)	MSE($\hat{u}$)
45	0.459	0.037	0.001	0.935	0.060	0.006
50	0.456	0.030	0.001	0.956	0.058	0.006
55	0.474	0.032	0.001	0.706	0.060	0.007

				$n = 200$				
m	$\hat{\sigma}$	SD($\hat{\sigma}$)	MSE($\hat{\sigma}$)	$\hat{D}_{s'}$	SD($\hat{D}_{s'}$)	MSE($\hat{D}_{s'}$)	E. MSE	F. MSE
45	1.77	0.195	0.055	0.151	0.012	0.001	0.063	0.912
50	1.80	0.124	0.051	0.096	0.016	0.001	0.059	0.908
55	1.51	0.197	0.057	0.142	0.015	0.001	0.066	1.117

TABLE 7.29
AR approximation with $d = 0.45$, $u = 0.8$, $D_{s'} = 0.1$, and iterations $= 1000$

			$n = 200$			
m	$\hat{d}$	SD($\hat{d}$)	MSE($\hat{d}$)	$\hat{u}$	SD($\hat{u}$)	MSE($\hat{u}$)
30	0.479	0.062	0.008	0.645	0.158	0.039
35	0.477	0.068	0.007	0.621	0.154	0.040
40	0.471	0.087	0.008	0.644	0.161	0.035

				$n = 200$				
m	$\hat{\sigma}$	SD($\hat{\sigma}$)	MSE($\hat{\sigma}$)	$\hat{D}_{s'}$	SD($\hat{D}_{s'}$)	MSE($\hat{D}_{s'}$)	E. MSE	F. MSE
30	1.38	0.123	0.075	0.119	0.051	0.007	0.129	0.990
35	1.39	0.118	0.065	0.118	0.053	0.007	0.119	0.974
40	1.41	0.125	0.077	0.118	0.083	0.008	0.128	0.986

TABLE 7.30
Optimal values of m with $u = 0.8$ using MA approximation and 1000 replications

n	$d=0.1/D_{s'}=0.45$	$d=0.3/D_{s'}=0.3$	$d=0.45/D_{s'}=0.1$
100	45	55	45
200	50	45	50

TABLE 7.31
Optimal values of m with $u = 0.8$ using AR approximation and 1000 replications

n	$d=0.1/D_{s'}=0.45$	$d=0.3/D_{s'}=0.3$	$d=0.45/D_{s'}=0.1$
100	35	45	35
200	40	45	35

7.11.2 Sunspots data

This application uses the sunspots data series considered by Tong (1990).
It displays periodicity with peaks and troughs as explained in Chapter 3

TABLE 7.32

Comparative assessment of optimal lag orders

	Optimal Lag Order – MA	Optimal Lag Order – AR
Series		
GARMA$(0, d, 0)$	[29,35]	[9,13]
GARSMA$(0,d,0)$x$(0,D_{s'},0)$	[45,55]	[35,45]

TABLE 7.33

MA approximation estimates for El' Nino series with $n = 1656$

Series	m	$\hat{d}$(MA)	$\hat{u}$(MA)	$\hat{D}_{s'}$(MA)
El' Nino	55	0.269(0.0009)	0.971(0.005)	0.324(0.0005)

illustrating seasonal cycles. The fitted optimal series model at $m = 45$ is $(1 - 2 \times 0.956B + B^2)^{0.43}(1 - B^{12})^{0.308} X_t = \epsilon_t$, with the standard errors for $\hat{d}$(MA), $\hat{u}$(MA), and $\hat{D}_{s'}$(MA) being 0.0004, 0.0018, and 0.0004, where $\epsilon_t \sim N(0, \sigma_\epsilon^2)$.

Remark 7.3. Therefore, from the empirical evidence, it is evident that one could only include toy applications for a real data case that depict periodicity.

7.12   Concluding Remarks

A comprehensive seasonal operator factorization and state space configuration for a GARSMA process are introduced. The methodology is employed to conduct Monte Carlo experiments for a small sample GARSMA model with a monthly seasonal periodicity. It returns an optimal lag order based on KF-based QML estimates validated by predictive accuracy. The AR approximation returns smaller lag order and MA representation provides better total MSE's for model profiles making it more feasible. A minor meta-analysis with GARMA model results of Chapter 3 was performed, and the better-approximating option was applied to a real application El' Nino series from environmental science. Probing the parent family of the introduced model in terms of persistence and stationarity through a unit root assessment becomes a worthwhile proposition. In line with the thesis outline, it is introduced in the next chapter titled as "Nearly efficient unit root tests for GARMA$(0, d, 0)$ processes with long memory".

7.13 Discussion

A truncated state space model entailing a KF was utilized to derive an efficient estimation/prediction framework for a long memory GARMA time series. It was used to generate QMLE estimates of the long memory parameter, noise and the Gegenbauer frequency index coefficient of various simulated series and applications belonging to the GARMA$(0, d, 0)$ process family. Within the same conceptual and methodological environment, multistep ahead forecasts were applied as diagnostic, prediction, and validating measures.

This exercise provides a truncated GARMA state space framework, an alternative estimation approach, a comparison of AR and MA approximation techniques in estimating parameters of a GARMA process and, most importantly, the introduction of an optimal lag truncation value (m) for estimation and forecasting as original contributions. Furthermore, the model presented in the chapter outperforms traditional time series forecasting models and closely matches extremely advanced modern hybrid models with neural networks. The utilization of the given conceptual paradigms is used later in the chapter to assess the features of seasonality in an extended GARSMA model, while an extension with realized volatility could form the basis for future research.

7.14 Chapter 7 Questions

1. Do GARMA models depict standard long memory or generalized long memory?

2. State the requisite criteria required in GARMA models to depict long memory?

3. What are the two main constituent equations of a state space model framework? Are they the same as for the ARFIMA model illustrated in the previous chapter?

4. Provide the system matrices of a GARMA state space configuration.

5. What requirements should a GARMA time series process possess in order to cast it in state space?

6. Name the two approximations that could be employed to estimate parameters of a long memory GARMA model using state space methodology? How does the linearization of the series taken care of to cast in state space?

7. Is it feasible and viable to use the quasi-maximum-likelihood presented in this chapter to provide estimates and predictions for GARMA parameters of a long-range-dependent series?

8. What is the KF? Explain in terms of the GARMA long memory time series?

9. What role does the KF play in estimation and prediction of parameters of a long memory GARMA series?

10. List the KF recursions required for the ARFIMA series in the previous chapter using the GARMA KF algorithm presented in this chapter?

8

Nonlinear and Nonstationary Time Series

Synopsis: Linear stochastic processes are said to have a unit root if 1 is a root of the process's characteristic equation. Such a process is nonstationary. If the other roots of the characteristic equation lie inside the unit circle (i.e., have a modulus or absolute value less than 1), then the first difference of the process will be stationary. Therefore, a *unit root* is a feature of processes that evolve through time, which can cause problems in statistical inference involving time series models.

8.1 Introduction

Fundamentally, in a unit root testing sense, a time series X_t is said to be integrated of order 1, if it becomes stationary after being differenced once. Then, we may write the null hypothesis of X_t as $I(1)$. On the contrary, a series that is stationary without being differenced is said to be integrated of order zero possessing a null hypothesis denoted as $I(0)$. Therefore, a series that becomes stationary after being differenced d times is said to be integrated of order d with a null hypothesis $I(d)$. Proceeding in this manner, it is often found that both the null hypotheses are rejected suggesting many series are not well represented by $I(0)$ and $I(1)$. In view of it, the class of fractionally integrated processes denoted as $FI(d)$, where order of integration d is extended to any real number has been introduced.

In a generic sense, unit root tests are consistent with a fairly low power if the alternative hypothesis is a $FI(d)$ process according to Diebold and Rudebusch (1991) and Lee and Schmidt (1996). This lack of power in general unit root tests has inspired and prompted researchers to create and adopt new testing procedures by taking the $FI(d)$ alternative into consideration. Therefore, unit root testing of GARMA$(0, d, 0)$ processes that fall into the extended $FI(d)$ class has aroused much interest.

In such a context, Dolado et al. (2002) extended a standard Dickey–Fuller (SDF) test to an augmented Dickey–Fuller (ADF) test for an ARFIMA model by employing the $FI(d)$ class. Yet, extending the methodology to assess a much more generalized GARMA model involves a multivariate dimension. Therefore, unit root testing of GARMA$(0, d, 0)$ processes for stationarity and long memory properties could be based on two parameters of the

DOI: 10.1201/9781032627007-8

model – either the long memory parameter or the degree of differencing d and/or polynomial index parameter u. Chung (1996) executed such a test for d and u using the conditional sum of squares (CSS) estimation method utilizing its concentrated profile likelihood function.

However, an extension of the unit root test based on state space methodology and KF estimates for a GARMA model is seemingly absent in the literature. Since the KF delivers QML estimates, an LR type test becomes the most feasible assessing option. Jansson and Nielsen (2012) proposed a nearly efficient unit root test that is applicable in such a context and assessing it's power becomes a worthwhile exercise.

The work presented in this chapter will comprise of two different tests using a nearly efficient variant of an LR type test. The first proposed test will revolve around u to check whether the desired series is a long memory GARMA$(0, d, 0)$ process or depicts standard long memory. Test to follow in this section is based on d to assess stationarity against nonstationarity.

Both these tests are based on state space modeling of long memory GARMA processes. They are done to assess properties of a GARMA$(0, d, 0)$ model revolving around QMLE estimators of u and d. In such a context, the QMLE estimators of u and d are calculated by maximizing ℓ_{σ^2} of Chapter 7. This method suggests that the development of the asymptotic theory shall hinge on the normality assumption. Therefore, an appropriate quasi-likelihood ratio unit root test based on Gaussian likelihood could be considered for a long memory GARMA$(0, d, 0)$ model. Jansson and Nielsen (2012) suggest the practical importance in exploring the power properties of their quasi-likelihood ratio test utilizing models with nuisance parameters and/or serial correlation. In that context as an extension to the established work of Jansson and Nielsen (2012), the construction of Jansson–Nielsen unit root tests in terms of u and d based on quasi-likelihood ratios (QLRs) are presented in the next section.

8.2 Construction of Jansson–Nielsen Type Tests for u and d

Consider the GARMA$(0, d, 0)$ process defined in Chapter 7 with ϵ_t as white noise.

It satisfies the criteria of Jansson–Nielsen type tests in the sense that it could be written in the form:

$$(1 - \nu B)\tau(B)X_t = \epsilon_t,$$

where $1 - \nu B = \frac{\phi(B)}{\theta(B)} = 1$ and $\tau(B) = (1 - 2uB + B^2)^d$, in accordance with equation (4) of Jansson and Nielsen (2012).

$$\text{Let}(1 - 2uB + B^2)^d = [2(u - 1)(1 - B) + (1 - B)^2 - 2(u - 1)]^d$$

$$= [2(u - 1)\Delta + \Delta^2 - 2(u - 1)]^d,$$

where $\Delta = (1 - B)$.

Hence, for a fixed d at $0.25 \le d < 0.5$, it becomes

$$[2(u-1)(\Delta-1) + \Delta^2]^d X_t = \epsilon_t.$$

Testing for u:

In order to test a null hypothesis of $H_0 : u = 1$ against an alternative hypothesis of $H_1 : u < 1$, we let $u = \phi + 1$ with $|u| < 1$ implying $\phi \in (-2, 0)$, to obtain the model:

$$[\Delta^2 + 2\phi(\Delta-1)]^d X_t = \epsilon_t, \tag{8.1}$$

Under the null hypothesis of $\phi = 0$ or $u = 1$, we get:

$$\Delta^{2d} X_t = \epsilon_t, \tag{8.2}$$

corresponding to X_t being a $FI(2d)$ process.

If $\phi < 0$, or $u < 1$, we get:

$$[\Delta^2 - 2\phi B]^d X_t = \epsilon_t. \tag{8.3}$$

Therefore, in order to formulate a test statistic in terms of the standard QLR testing, following $H_0 : u = 1$ against $H_1 : u < 1$ is equivalent to the test:

$$H_0 : \phi = 0. \tag{8.4}$$

$$H_1 : \phi < 0. \tag{8.5}$$

Note: This test will assess if the model depicts standard long memory or a GARMA process with long memory.

Since a QLR test has been formulated in terms of the parameter u, a quasi-likelihood ratio-type test statistic based in line with the result of Jansson and Nielsen (2012) will have the form:

$$LR_n^u = \max_{\bar{u} \le 1}[\ell_{\sigma^2}(d, \bar{u})] - \max[\ell_{\sigma^2}(d, 1)], \tag{8.6}$$

where $\bar{u}$ is the mean of u within the estimation profile domain and $\ell_{\sigma^2}(\bullet, \bullet)$ is from Chapter 7.

Testing for d:

Consider the GARMA$(0, d, 0)$ process model defined in Chapter 7.

It's Gegenbauer polynomial raised to the power d could be written as:

$$(1 - 2uB + B^2)^d = (1 - 2uB + B^2)^{d'-1/2}, \tag{8.7}$$

with conditions $|u| < 1$, $d' \in (0.5, 1.0)$ and $d = d' - 1/2$.

Therefore, in order to formulate a test statistic in terms of a standard QLR test, following hypotheses could be proposed:

Null hypothesis

$$H_0 : d' = 1, \tag{8.8}$$

against the alternative of

$$H_1 : d' < 1. \tag{8.9}$$

Since a QLR test has been formulated in terms of the parameter d', a quasi-likelihood ratio-type test statistic based in line with the result of Jansson and Nielsen (2012) will have the form:

$$LR_n^{d'} = \max_{\bar{d}' \leq 1}[\ell_{\sigma^2}(d' - 1/2, u)] - \max[\ell_{\sigma^2}(1, u)], \tag{8.10}$$

where $\bar{d}'$ is the mean of d' within the estimation profile domain and $\ell_{\sigma^2}(\bullet, \bullet)$ is from Chapter 7.

This test will assess if the model is nonstationary or stationary for a given parameter profile. Asymptotics related to the state space modeling based estimation of long memory Gegenbauer processes in creating the Jansson–Nielsen QLR tests are presented in the next section based on the work of Chan and Palma (1998).

8.3   Asymptotic Properties of State Space-Based QMLE Estimation

Initially, prior to stating the main theorems, we introduce some definitions, regularity conditions, and notation. Let $\hat{\Theta}_{n,m} = (\hat{d}, \hat{u}, \hat{\sigma})'$ be the QMLE estimator that maximizes the approximate quasi log-likelihood of a truncated GARMA$(0, d, 0)$ model such that $\Theta = (d, u, \sigma_\epsilon)' = (\Theta_{01}, \Theta_{02}, \Theta_{03})'$ is a three-dimensional true parameter estimate vector. Assume that the regularity conditions in Dahlhaus (1989) hold.

Let the partial derivatives be defined as:

$$\nabla f(\Theta) = (\frac{\partial}{\partial \Theta_j} f(\Theta))_{j=1,...,r}$$

and

$$\nabla^2 f(\Theta) = (\frac{\partial^2}{\partial \Theta_j \partial \Theta_k} f(\Theta))_{j,k=1,...,r},$$

and the matrices for $i, j = 1, ..., r$ are defined as:

$$T_{\partial i} = T(\frac{\partial}{\partial \Theta_i} f_\Theta), \quad T_{\partial i \partial j} = T(\frac{\partial^2}{\partial \Theta_i \partial \Theta_j} f_\Theta),$$

$$A_i^{(1)} = T^{-1} T_{\partial i} T^{-1}, \quad A^{(1)} = T^{-1} T_\nabla T^{-1},$$

$$\hat{A}^{(1)} = \hat{T}^{-1}\hat{T}_\nabla \hat{T}^{-1}, \quad \hat{A}_i^{(1)} = \hat{T}^{-1}\hat{T}_{\partial i}\hat{T}^{-1},$$

$$A_{ij}^{(2)} = T^{-1}T_{\partial i}T^{-1}T_{\partial j}T^{-1}, \quad A^{(2)} = T^{-1}T_\nabla T^{-1}T_\nabla T^{-1},$$

$$\hat{A}^{(2)} = \hat{T}^{-1}\hat{T}_\nabla \hat{T}^{-1}\hat{T}_\nabla \hat{T}^{-1}, \quad \hat{A}_{ij}^{(2)} = \hat{T}^{-1}\hat{T}_{\partial i}\hat{T}^{-1}\hat{T}_{\partial j}\hat{T}^{-1},$$

$$A_{ij}^{(3)} = T^{-1}T_{\partial i\partial j}T^{-1}, \quad A^{(3)} = T^{-1}T_{\nabla^2}T^{-1},$$

$$\hat{A}^{(3)} = \hat{T}^{-1}\hat{T}_{\nabla^2}\hat{T}^{-1}, \quad \hat{A}_{ij}^{(3)} = \hat{T}^{-1}\hat{T}_{\partial i\partial j}\hat{T}^{-1},$$

with $T_n(\Theta)$ being the covariance matrix of $(X_{1,m}, X_{2,m}, ..., X_{n,m})'$ for series length (n) such that $T = T_n(f_\Theta)$, $T_\nabla = T_n(\nabla f_\Theta)$, $T_{\nabla^2} = T_n(\nabla^2 f_\Theta)$, $\hat{T} = T_n(\hat{f}_{\Theta,m})$, $\hat{T}_\nabla = T_n(\nabla \hat{f}_{\Theta,m})$, $\hat{T}_{\nabla^2} = T_n(\nabla^2 \hat{f}_{\Theta,m})$, and $f(\bullet)$ denoting the spectrum defined in (3.2).

The three results shown next establish consistency, asymptotic normality, and efficiency of truncated QMLE's in terms of GARMA$(0, d, 0)$ model parameters.

Theorem 8.1 (Consistency): Assume that $m = n^\beta$ with $\beta > 0$, then as $n \to \infty$,
$$\hat{\Theta}_{n,m} \to \Theta \quad \text{in probability.}$$

Theorem 8.2 (Central Limit Theorem): Suppose that $m = n^\beta$ with $\beta > 1/2$, then as $n \to \infty$,

$$\sqrt{(n)}(\hat{\Theta}_{n,m} - \Theta) \to_\mathcal{L} N(0, \Sigma(\Theta)),$$

where "$\to_\mathcal{L}$" denotes convergence in distribution and $\Sigma^{-1}(\Theta) = (\Sigma_{ij}^{-1}(\Theta))$ with

$$\Sigma_{ij}^{-1}(\Theta) = \frac{1}{4\pi} \int_{-\pi}^{\pi} \{\frac{\partial logk(\omega, \Theta)}{\partial \Theta_i}\}\{\frac{\partial logk(\omega, \Theta)}{\partial \Theta_j}\} d\omega,$$

and

$$k(\omega, \Theta) = |\sum_{j=0}^{\infty} \psi_j(\Theta)e^{ij\omega}|^2,$$

where ω is defined as *prediction error covariance estimate*.

Theorem 8.3 (Efficiency): Assume that $m = n^\beta$ with $\beta > 1/2$, then $\hat{\Theta}_{n,m}$ is an *efficient estimator* of Θ_0.

Note: Next the proofs of Theorems 8.1–8.3 are provided.

Proofs of Theorems 8.1–8.3

Prior to proving Theorems 8.1–8.3, certain auxiliary lemmas need to be established. It is done next with the help of a generic constant K.

Lemma 8.1. Let C_j be the coefficients given earlier, then for j large and $u > 0$,

$$|C_j(\Theta)| \leq Kj^{d-2}.$$

Proof. $C(z) = (\vartheta(z)/\Phi(z))(1-z)^{-d+1} = \sum_{j=0}^{\infty} C_j z^j$. Define $\varphi(z) = (\vartheta(z)/\Phi(z)) = \sum_{k=0}^{\infty} \varphi_k z^k$ and $\zeta(z) = (1-z)^{-d+1} = \sum_{k=0}^{\infty} \zeta_k z^k$. The coefficients φ_k can be written [see page 92 of Brockwell and Davis (1991)] as:

$$\varphi_k = \sum_{i=1}^{m} \sum_{l=0}^{r_i-1} \alpha_{il} k^l \varepsilon_i^k, \quad k \geq \max(p, q+1) - p,$$

where m is the number of distinct roots of $\Phi(z), |\varepsilon_i| < 1$ is the inverse of the ith root of $\Phi(z)$ with multiplicity, and r_i and α_{il} are constants. Let $L \geq \max(p, q+1) - p$, then

$$|C_j| = |\sum_{k=0}^{L} \varphi_k \zeta_{j-k} + \sum_{k=L+1}^{\infty} \varphi_k \zeta_{j-k}|$$

$$= |\sum_{k=0}^{L} \varphi_k \zeta_{j-k} + \sum_{k=L+1}^{\infty} [\sum_{i=1}^{m} \sum_{l=0}^{r_i-1} \alpha_{il} k^l \varepsilon_i^k] \zeta_{j-k}|$$

$$\leq \sum_{k=0}^{L} |\varphi_k||\zeta_{j-k}| + \sum_{i=1}^{m} \sum_{l=0}^{r_i-1} |\alpha_{il}|[\sum_{k=L+1}^{\infty} k^l |\varepsilon_i|^k]|\zeta_{j-k}|].$$

$\square$

Note: For $k \geq L$, there exists constants $c_l \geq 0$ and $0 < a_i < 1$ such that $k^l|\varepsilon_i|^k \leq c_l a_i^k$. This can be seen as follows. Let $a_i = |\varepsilon_i| + u_i$, where $u_i > 0$. Then, $0 < a_i < 1$ (since $|\varepsilon_i| < 1$, it is always possible to find such an ϵ_i). For $l \geq 0$, $k^l(|\varepsilon_i|/a_i)^k \to 0$ as $k \to \infty$. Thus, for $k \geq L$, there is a constant c_l such that $k^l(|\varepsilon_i|/a_i)^k \leq c_l$, which implies $k^l|\varepsilon_i|^k \leq c_l a_i^k$. Therefore,

$$|C_j| \leq \sum_{k=0}^{L} |\varphi_k||\zeta_{j-k}| + \sum_{i=1}^{m} \sum_{l=0}^{r_i-1} |\alpha_{il}|[\sum_{k=L+1}^{\infty} c_l a_i^k |\zeta_{j-k}|]$$

$$\leq \sum_{k=0}^{L} |\varphi_k||\zeta_{j-k}| + \sum_{i=1}^{m} \sum_{l=0}^{r_i-1} |\alpha_{il}|c_l[\sum_{k=L+1}^{\infty} a_i^k |\zeta_{j-k}|]$$

$$= \sum_{k=0}^{L} |\varphi_k||\zeta_{j-k}| + |\zeta_j| \sum_{i=1}^{m} \sum_{l=0}^{r_i-1} |\alpha_{il}|c_l[\sum_{k=L+1}^{\infty} a_i^k |P(k,j,d)|],$$

$$\text{where} \quad P(k,j,d) = \frac{\zeta_{j-k}}{\zeta_j} = \frac{j!(j-k+d-2)!}{(j-k)!(j+d-2)!},$$

$$\text{since} \zeta_{j-k} = \frac{(j-k+d-2)!}{(j-k)!(d-2)!} \text{ and } \zeta_j = \frac{(j+d-2)!}{j!(d-2)!}.$$

Thus,

$$|C_j| \le \sum_{k=0}^{L} |\varphi_k||\zeta_{j-k}| + |\zeta_j| \sum_{i=1}^{m} \sum_{l=0}^{r_i-1} |\alpha_{il}||c_l| F(1,-j,2-d-j,a_i),$$

where F is the hypergeometric function [see Hosking (1981)]. Moreover, $|\zeta_{j-k}| \le |\zeta_{j-L}|$, for $0 \le k \le L$. This implies

$$|C_j| \le |\zeta_{j-L}| \sum_{k=0}^{L} |\varphi_k| + |\zeta_j| \sum_{i=1}^{m} \sum_{l=0}^{r_i-1} |\alpha_{il}||c_l| F(1,-j,2-d-j,a_i).$$

As $j \to \infty$, $F(1,-j,2-d-j,a_i) \to (1-a_i)^{-1}$ [see Hosking (1981)], and for j large, $|\zeta_j| \sim Kj^{d-2}$ and $|\zeta_{j-L}| \sim Kj^{d-2}$ (for fixed L). Therefore, for large j,

$$|C_j| \le Kj^{d-2} [\sum_{k=0}^{L} |\varphi_k| + \sum_{i=1}^{m} \sum_{l=0}^{r_i-1} \frac{|\alpha_{il}|}{1-a_i}].$$

Since $[\sum_{k=0}^{L} |\varphi_k| + \sum_{i=1}^{m} \sum_{l=0}^{r_i-1} \frac{|\alpha_{il}|}{1-a_i}]$ is a rational function of the parameter $\Theta \in \vartheta$, it is continuous. Given that the parameter space ϑ is compact, there is a constant K such that

$$\sum_{k=0}^{L} |\varphi_k| + \sum_{i=1}^{m} \sum_{l=0}^{r_i-1} \frac{|\alpha_{il}|}{1-a_i} \le K.$$

Thus,

$$|C_j| \le Kj^{d-2},$$

and it completes the proof of Lemma 8.1.

Lemma 8.2. If $\{a_k\}$ is a sequence of numbers, then for large n,

$$\frac{\sum_{k=n}^{\infty} k^{2d-4}}{n^{2d-3}} \le K.$$

Proof. Let $\beta = 2d - 3 < 0$. Then

$$\lim_{n \to \infty} \frac{\sum_{k=n}^{\infty} k^{\beta-1}}{n^{\beta}} = \lim_{n \to \infty} \sum_{k=n}^{\infty} (k/n)^{\beta-1} \frac{1}{n}.$$

$$= \lim_{n \to \infty} \sum_{j=0}^{\infty} (j/n+1)^{\beta-1} \frac{1}{n}.$$

By Polya and Szego [(1992), page 53], the last sum equals $\int_0^{\infty} (x+1)^{\beta-1}dx = 1/|\beta|$ and it completes the proof. $\square$

Lemma 8.3. Let $m = n^\beta$ with $\beta > 0$. Then, as $n \to \infty$ uniformly in Θ, we have the following:

(i) For $\beta > 0, \|T^{-1} - \hat{T}^{-1}\| \to 0$,

(ii) for $\beta > 1/2, \sqrt{(n)}\|A_i^{(1)} - \hat{A}_i^{(1)}\| \to 0$, for $i = 1, ..., r$,

(iii) for $\beta > 1/2, \|A_{ij}^{(2)} - \hat{A}_{ij}^{(2)}\| \to 0$, for $i, j = 1, ..., r$,

(iv) for $\beta > 1/2, \|A_{ij}^{(3)} - \hat{A}_{ij}^{(3)}\| \to 0$, for $i, j = 1, ..., r$.

Proof. For (i), the matrix T satisfies $x'Tx \geq Kx'x$ uniformly in Θ, where K is a constant. From Lemma 8.1, it is evident that for $u > 0$, there exists n_0, independent of Θ, such that for $n \geq n_0$, $\|T - \hat{T}\| \leq u$, where $\|A\|^2 = \sup_{x \in R^n}((x'AA'x)/x'x)$ is the square of the spectral norm of the matrix A. Thus, $|x'Tx - x'\hat{T}x| \leq ux'x$, and $x'\hat{T}x \geq x'Tx - ux'x \geq (k-u)x'x$. $\square$

It follows that $\|\hat{T}^{-1}\| \leq K$ uniformly in Θ. Using the property $\|AB\| \leq \|A\|\|B\|$ [see Dahlhaus (1989)], $\|T^{-1} - \hat{T}^{-1}\| \leq \|T^{-1}\|\|\hat{T}^{-1}\|\|T - \hat{T}\| \leq K\|T - \hat{T}\|$. By using the theory in Lemma 8.1, it is evident that the last term tends to zero uniformly in Θ and hence completes the proof of (i).

For (ii), we observe that

$$\sqrt{(n)}\|A_i^{(1)} - \hat{A}_i^{(1)}\| = \sqrt{(n)}\|T^{-1}T_{\partial i}T^{-1} - \hat{T}^{-1}\hat{T}_{\partial i}\hat{T}^{-1}\|$$

$$\leq \sqrt{(n)}\|T^{-1}T_{\partial i}T^{-1} - T^{-1}T_{\partial i}\hat{T}^{-1}\|$$

$$+\sqrt{(n)}\|T^{-1}T_{\partial i}\hat{T}^{-1} - \hat{T}^{-1}T_{\partial i}\hat{T}^{-1}\|$$

$$+\sqrt{(n)}\|\hat{T}^{-1}(T_{\partial i} - \hat{T}_{\partial i})\hat{T}^{-1}\|$$

$$\leq \sqrt{(n)}\|T^{-1}T_{\partial i}\|\|T^{-1} - \hat{T}^{-1}\|$$

$$+\sqrt{(n)}\|\hat{T}^{-1}T_{\partial i}\|\|T^{-1} - \hat{T}^{-1}\| + \sqrt{(n)}\|\hat{T}^{-1}\|^2\|T_{\partial i} - \hat{T}_{\partial i}\|$$

$$\leq K\sqrt{(n)}\|T^{-1} - \hat{T}^{-1}\| + K\sqrt{(n)}\|T_{\partial i} - \hat{T}_{\partial i}\|$$

$$\leq Kn^{(2d-2)\beta+1/2}.$$

Therefore, for $0 < d < 1/2$ and $\beta \geq 1/2$, uniform convergence is obtained completing the proof for (ii). The proofs of parts (iii) and (iv) are analogous to the proof of (ii).

Lemma 8.4. For $\beta \geq 1/2$, as $n \to \infty$, uniformly in Θ,

(i) $\frac{1}{\sqrt{(n)}}|tr[T^{-1}T_{\partial i} - \hat{T}^{-1}\hat{T}_{\partial i}]| \to 0$ for $i = 1, ..., r$,

(ii) $\frac{1}{n}|tr[T^{-1}T_{\partial i}T^{-1}T_{\partial j} - \hat{T}^{-1}\hat{T}_{\partial i}\hat{T}^{-1}\hat{T}_{\partial j}]| \to 0$ for $i, j = 1, ..., r$,

(iii) $\frac{1}{\sqrt{(n)}}|tr[T^{-1}T_{\partial i\partial j} - \hat{T}^{-1}\hat{T}_{\partial i\partial j}]| \to 0$ for $i, j = 1, ..., r$.

Proof. For (i), we observe that

$$\frac{1}{\sqrt{(n)}}|tr[T^{-1}T_{\partial i} - \hat{T}^{-1}\hat{T}_{\partial i}]| \leq \frac{1}{\sqrt{(n)}}|tr[T^{-1}T_{\partial i} - T^{-1}\hat{T}_{\partial i}]|$$

$$+ \frac{1}{\sqrt{(n)}}|tr[(T^{-1} - \hat{T}^{-1})\hat{T}_{\partial i}]|$$

$$\leq \frac{1}{\sqrt{(n)}}||T^{-1}||||T_{\partial i} - \hat{T}_{\partial i}|| + \frac{1}{\sqrt{(n)}}||T^{-1} - \hat{T}^{-1}||||\hat{T}_{\partial i}||$$

$$\leq K\sqrt{(n)}||T_{\partial i} - \hat{T}_{\partial i}]|| + K\sqrt{(n)}||T^{-1} - \hat{T}^{-1}||$$

$$\leq Kn^{(2d-2)\beta+1/2}. \qquad \square$$

Therefore, from an extension of Lemma 8.3(i) for $0 < d < 1/2$ and $\beta \geq 1/2$, Lemma 8.4(i) holds. The proofs of parts 8.4(ii) and 8.4(iii) are analogous to proof of 8.4(i).

Lemma 8.5. Let $m = n^\beta$ with $\beta > 0$. Then, uniformly in Θ,
$\lim_{n\to\infty} \frac{1}{n}log[det\{\hat{T}T^{-1}\}] = 0$

Proof.

$$\frac{1}{n}log[det\{\hat{T}T^{-1}\}] \leq log\{\frac{1}{n}tr\{TT^{-1}\}\} = log\{\frac{1}{n}tr\{T^{-1}[\hat{T} - T]\} + 1\}.$$

Since $T^{-1} \leq KI_n$ uniformly in Θ,

$$|\{\frac{1}{n}tr\{T^{-1}[\hat{T} - T]\}\}| \leq K|\{\frac{1}{n}tr\{\hat{T} - T\}\}| = K\sum_{k=m+1}^{\infty} C_k^2(\Theta). \qquad \square$$

From Lemma 8.1, $\sum_{k=m+1}^{\infty} k^{2d-4} \leq Km^{2d-3}$. Therefore, $\sum_{k=m+1}^{\infty} C_k^2(\Theta) \leq Km^{2d-3} = Kn^{\beta(2d-3)}$. Since $2d - 3 < 0$ and $\beta > 0$, $n^{\beta(2d-3)} \to 0$ as $n \to \infty$. Thus, it completes the proof.

Lemma 8.6. Let $m = n^\beta$ with $\beta > 0$. Then, as $n \to \infty$,

$$\sup_{\Theta} |\hat{L}_n(\Theta) - L_n(\Theta)| \to 0 \quad a.s.$$

Proof. Observe that

$$|\hat{L}_n(\Theta) - L_n(\Theta)| \le \frac{1}{2n} log[det\{\hat{T}T^{-1}\}] + \frac{1}{2n} Z'_n Z_n ||\hat{T}^{-1} - T^{-1}||.$$

$\square$

From Lemma 8.5, $\frac{1}{2n} log[det\{\hat{T}T^{-1}\}] \to 0$ and from Hannan(1979, page 133), $\frac{1}{2n} Z'_n Z_n \to \gamma_0/2$ a.s. as $n \to \infty$, where γ_0 is the variance of the process $\{Z_t\}$. From Lemma 8.3(i), $||\hat{T}^{-1} - T^{-1}|| \to 0$ uniformly in Θ as $n \to \infty$. It completes the proof of our result.

Proof of Theorem 8.1 (Consistency): From Lemma 8.6, $\sup_\Theta |\hat{L}_n(\Theta) - L_n(\Theta)| \to 0$ a.s., as $n \to \infty$. Therefore, clearly

$$-\hat{L}_n(\Theta) \le |\hat{L}_n(\Theta) - L_n(\Theta)| - L_n(\Theta).$$

Then,

$$\sup_\Theta \{-\hat{L}_n(\Theta)\} \le \sup_\Theta |\hat{L}_n(\Theta) - L_n(\Theta)| + \sup_\Theta \{-L_n(\Theta)\}.$$

Equivalently,

$$-\inf_\Theta \hat{L}_n(\Theta) \le \sup_\Theta |\hat{L}_n(\Theta) - L_n(\Theta)| - \inf_\Theta L_n(\Theta),$$

or

$$\inf_\Theta L_n(\Theta) - \inf_\Theta \hat{L}_n(\Theta) \le \sup_\Theta |\hat{L}_n(\Theta) - L_n(\Theta)|.$$

Similarly,

$$\inf_\Theta \hat{L}_n(\Theta) - \inf_\Theta L_n(\Theta) \le \sup_\Theta |\hat{L}_n(\Theta) - L_n(\Theta)|.$$

Thus,

$$|\inf_\Theta \hat{L}_n(\Theta) - \inf_\Theta L_n(\Theta)| \le \sup_\Theta |\hat{L}_n(\Theta) - L_n(\Theta)|.$$

Therefore,

$$|\inf_\Theta \hat{L}_n(\Theta) - \inf_\Theta L_n(\Theta)| \to 0 \quad a.s., \quad as \quad n \to \infty. \qquad (8.11)$$

Let $U(\Theta_0)$ be a neighborhood of Θ_0 with radius d_0. Using the triangle inequality, we get

$$|\inf_\Theta \hat{L}_n(\Theta) - \inf_{U(\Theta_0)} L_n(\Theta)| \le |\inf_\Theta \hat{L}_n(\Theta) - \inf_\Theta L_n(\Theta)| + |\inf_\Theta L_n(\Theta) - \inf_{U(\Theta_0)} L_n(\Theta)|.$$

Based on equation (8.11), first term on the right-hand side converges to zero a.s., and hence in probability. On the other hand, $|\inf_\Theta L_n(\Theta) - \inf_{U(\Theta_0)} L_n(\Theta)| \to 0$, in probability, since the QMLE, $\hat{\Theta}_n \to \Theta_0$ in probability, as per Theorem 3.1 of Dahlhaus (1989). Thus, $|\inf_\Theta \hat{L}_n(\Theta) - \inf_{U(\Theta_0)} L_n(\Theta)| \to 0$ in probability, implying $\Theta_{n,m} \to \Theta_0$ in probability as required.

Proof of Theorem 8.2 (Central Limit Theorem): It suffices to prove that:

(i) $\sup_{\Theta \in \vartheta} \sqrt{(n)} |\nabla L_n(\Theta) - \nabla \hat{L}_n(\Theta)| \to 0$ a.s.,

(ii) $\sup_{\Theta \in \vartheta} |\nabla^2 L_n(\Theta) - \nabla^2 \hat{L}_n(\Theta)| \to 0$ a.s.,

(i) The gradient of $L_n(\Theta)$ can be written as:

$$\nabla L_n(\Theta) = \frac{1}{2n} tr[T^{-1} T_\nabla] - \frac{1}{2n} Z_n' A^{(1)} Z_n$$

and

$$\nabla \hat{L}_n(\Theta) = \frac{1}{2n} tr[\hat{T}^{-1} \hat{T}_\nabla] - \frac{1}{2n} Z_n' \hat{A}^{(1)} Z_n.$$

Hence,

$$\sqrt{(n)} |\nabla L_n(\Theta) - \nabla \hat{L}_n(\Theta)| \le \frac{1}{2\sqrt{(n)}} |tr[T^{-1} T_\nabla - \hat{T}^{-1} \hat{T}_\nabla]|$$

$$+ \frac{Z_n' Z_n}{2n} \sqrt{(n)} ||A^{(1)} - \hat{A}^{(1)}||.$$

Observe that

$$\frac{1}{2\sqrt{(n)}} |tr[T^{-1} T_\nabla - \hat{T}^{-1} \hat{T}_\nabla]| = \frac{1}{2\sqrt{(n)}} (\sum_{i=1}^{r} |tr[T^{-1} T_{\partial i} - \hat{T}^{-1} \hat{T}_{\partial i}]|^2)^{1/2}.$$

From Lemma 8.4(i), $\frac{1}{2\sqrt{(n)}} |tr[T^{-1} T_{\partial i}] - tr[\hat{T}^{-1} \hat{T}_{\partial i}]|$ goes to zero uniformly in Θ, as $n \to \infty$. Hence, the same result holds for $\frac{1}{2\sqrt{(n)}} |tr[T^{-1} T_\nabla - tr \hat{T}^{-1} \hat{T}_\nabla]|$.

On the other hand, $\frac{Z_n' Z_n}{2n}$ is asymptotically equal to $\gamma_0/2$ and $\sqrt{(n)} ||A^{(1)} - \hat{A}^{(1)}|| = \sqrt{(n)} (\sum_{i=1}^{r} ||A_i^{(1)} - \hat{A}_i^{(1)}||^2)^{1/2}.$

By Lemma 8.3(ii), for $i = 1, ..., r$, $\sqrt{(n)} ||A_i^{(1)} - \hat{A}_i^{(1)}|| \to 0$ as $n \to \infty$ uniformly in Θ. Thus, (i) of Theorem 8.2 is established.

(ii) The second derivatives of $L_n(\Theta)$ can be written as

$$\nabla^2 L_n(\Theta) = -\frac{1}{2n} tr[T^{-1} T_\nabla T^{-1} T_\nabla] + \frac{1}{2n} tr[T^{-1} T_{\nabla^2}]$$

$$+ \frac{1}{2n} X_n' A^{(2)} X_n - \frac{1}{2n} Z_n' A^{(3)} Z_n$$

and

$$\nabla^2 \hat{L}_n(\Theta) = -\frac{1}{2n} tr[\hat{T}^{-1} \hat{T}_\nabla \hat{T}^{-1} \hat{T}_\nabla] + \frac{1}{2n} tr[\hat{T}^{-1} \hat{T}_{\nabla^2}]$$

$$+ \frac{1}{2n} X_n' \hat{A}^{(2)} X_n - \frac{1}{2n} Z_n' \hat{A}^{(3)} Z_n$$

Hence,

$$|\nabla^2 L(\Theta) - \nabla^2 \hat{L}(\Theta)| \leq \frac{1}{2n}|tr\{T^{-1}T_\nabla T^{-1}T_\nabla - \hat{T}^{-1}\hat{T}_\nabla \hat{T}^{-1}\hat{T}_\nabla\}|$$

$$+\frac{1}{2n}|tr\{T^{-1}T_{\nabla^2} - \hat{T}^{-1}\hat{T}_{\nabla^2}\}|$$

$$+\frac{1}{2n}X'_n X_n ||A^{(2)} - \hat{A}^{(2)}|| + \frac{1}{2n}Z'_n Z_n ||A^{(3)} - \hat{A}^{(3)}||$$

$$= I + II + III, \quad say.$$

Now

$$I = \frac{1}{2}(\sum_{i=1}^{r}[\frac{1}{n}|tr\{T^{-1}T_{\partial i}T^{-1}T_{\partial i} - \hat{T}^{-1}\hat{T}_{\partial i}\hat{T}^{-1}\hat{T}_{\partial i}\}|]^2)^{1/2}.$$

By Lemma 8.4(ii), this term converges to zero for $\beta \geq 1/2$ uniformly in Θ. Also,

$$II = \frac{1}{2n}|\sum_{i=1}^{r}tr\{T^{-1}T_{\partial i\partial i} - \hat{T}^{-1}\hat{T}_{\partial i\partial i}\}|$$

$$\leq \frac{1}{2}\sum_{i=1}^{r}\frac{1}{n}|tr\{T^{-1}T_{\partial i\partial i} - \hat{T}^{-1}\hat{T}_{\partial i\partial i}\}|,$$

and by virtue of Lemma 8.4(iii), the right-hand side converges to zero for $\beta \geq 1/2$ uniformly in Θ. Finally, since $Z'_n Z_n/2n \sim \gamma_0/2$, $||A^{(2)} - \hat{A}^{(2)}|| = (\sum_{i,j=1}^{r}||A_{ij}^{(2)} - \hat{A}_{ij}^{(2)}||^2)^{1/2} \to 0$ as $n \to \infty$ from Lemma 8.3(iii), and $||A^{(3)} - \hat{A}^{(3)}|| = (\sum_{i,j=1}^{r}||A_{ij}^{(3)} - \hat{A}_{ij}^{(3)}||^2)^{1/2} \to 0$ according to Lemma 8.3(iv), $III \to 0$, and hence, (ii) is established completing the proof.

Proof of Theorem 8.3 (Efficiency): The proof is based on the Fisher information matrices $I_n(\Theta_0)$ and $\hat{I}_n(\Theta_0)$ of the exact and truncated log likelihood models evaluated at Θ_0, respectively.

Therefore,

$$(1/n)I_n(\theta_0) - (1/n)\hat{I}_n(\theta_0) = ntr\{T^{-1}T_\nabla T^{-1}T_\nabla - \hat{T}^{-1}\hat{T}_\nabla \hat{T}^{-1}\hat{T}_\nabla\}.$$

where

$$T = T_n(f)$$

and

$$T_\nabla = T_n(\nabla f)$$

with f being the spectral density of the process.
Applying Lemma 8.4(ii) yields

$$\lim_{n\to\infty}\{(1/n)I_n(\theta_0 - (1/n)\hat{I}_n(\theta_0\} = 0.$$

Finally, applying Theorem 5.1 of Dahlhaus (1989),

$$\lim_{n \to \infty} (1/n)\hat{I}_n(\theta_0) = \Gamma(\theta_0),$$

completing the proof.

Assessing the strength of the above tests in terms of their power is a worthwhile exercise. Especially, since it could provide a remedy to the perception that general unit root tests with an $FI(d)$ alternative hypothesis lacks power. It becomes the topic of interest in the next section.

8.4   Power of Tests

Monte Carlo experiments were done to assess the *power* of the tests introduced for varying combinations of parameters. The results are illustrated in Tables 8.1–8.2.

Note: It is clear that the notion of a general unit root test with $FI(d)$ alternative hypothesis lacking power as mentioned by Hassler and Wolters (1994), Diebold et al. (1991), and Lee and Schmidt (1996) is addressed to a great extent by the tests discussed in this chapter with minimal hassle. The values obtained for the power of the tests are reasonable and comparable with similar results for the same model in Beaumont and Ramachandran (2001). Furthermore, in a practical sense, adequately powerful tests introduced in this chapter

TABLE 8.1
The power of testing for u with 1000 replications for $H_0 : u = 1$ and $H_1 : u < 1$

Power					
n	$(d = 0.1)$	$(d = 0.2)$	$(d = 0.3)$	$(d = 0.4)$	$(d = 0.45)$
100	71.9%	74.5%	79.1%	79.7%	79.9%
200	81.3%	84.8%	89.2%	89.3%	89.8%

TABLE 8.2
The power of testing for d with 1000 replications for $H_0 : d' = 1$ and $H_1 : d' < 1$

Power					
n	$(d' = 0.6)$	$(d' = 0.7)$	$(d' = 0.8)$	$(d' = 0.9)$	$(d' = 0.95)$
100	50.8%	69.5%	73.8%	77.2%	79.7%
200	70.8%	74.3%	82.6%	83.1%	83.9%

will be appropriate and feasible tools to employ on potential applications such as "international interest rates". In such a context, a GARMA$(0, d, 0)$ model would be a best fit as shown by Beaumont and Ramachandran (2001). For testing the fractional order of seasonal and nonseasonal unit roots of long memory processes, see Ferrara et al. (2010).

By considering all the facts given in the chapter, concluding remarks are provided in the next section.

8.5 Discussion

A "nearly efficient unit root testing procedure" for QMLEs derived from a state space model of a long memory Gegenbauer process is introduced. It is utilized as a dual purpose mechanism to assess stationarity versus nonstationarity and standard long memory versus long memory Gegenbauer process attributes. Asymptotic properties of the QMLE estimator of a long memory GARMA$(0, d, 0)$ process is provided as an additional contribution. The efficiency of the test is corroborated by the Monte Carlo evidence of its power, in refuting an established notion of generic unit root tests with fractionally integrated alternative hypotheses are consistent with a fairly low power.

8.6 Chapter 8 Questions

1. What is a linear time series?

2. What is a nonlinear time series?

3. What is a stationary time series?

4. What is a nonstationary time series?

5. What is a unit root?

6. Pick a GARMA (0,d,0) series of your choice (different to the ones presented in this chapter) and conduct the hypothesis test for parameter u to assess if the model can be classified as a standard or generalized long memory process?

7. Pick a GARMA (0,d,0) series of your choice (different to the ones presented in this chapter) and conduct the hypothesis test for parameter d to assess if the model can be classified as a stationary or a nonstationary process?

8. Try to run the test done in question 6 by writing a short programming code in R?

9. Try to run the test done in question 7 by writing a short programming code in R?

10. What is a unit root test?

11. The most popular unit root test is the augmented Dickey–Fuller (ADF) test. To execute such a test in R, run the following code and interpret the results. Place the hash sign in front of the comments below while running the program in R. The comments are provided as documentation and are not required as syntax.

adf ¡- data.frame(k = 0:9,

level = NA,

$diff_1$ = NA,

$diff_2$ = NA)

Comment: Run tests for a series of models

for (i   in   1:nrow(adf))

k ¡- adf k[i]

pos ¡- (9 - k + 1):nrow(data) Position of used observations

for (j in c("level", "diff₁", "diff₂"))

adf_test ¡- adf_test(data[pos, j], alternative = "stationary", k = k)

adf[i, j] ¡- adf_test p.value

Comment: Show results

adf

12. Run the following elementary code in R on unit root testing and interpret the results.

Comment: Time Series

Comment: A time series which contains no unit-root:

x = rnorm(1000)

Comment: A time series which contains a unit-root:

y = cumsum(c(0, x))

Comment: adfTest -

adfTest(x)

adfTest(y)

Comment: unitrootTest -

unitrootTest(x)

unitrootTest(y)

9

An Introduction to Nonparametric Long Memory Time Series

Synopsis: Time series analysis has traditionally relied on parametric models that assume specific functional forms for the underlying data-generating process. While these models, such as autoregressive (AR), moving average (MA), and their combinations (ARMA, ARIMA), have proven invaluable in many applications, they often impose restrictive assumptions that may not hold in practice. When the true structure of the time series is unknown or when the data exhibit complex, nonlinear behaviors, parametric approaches may fail to capture important features of the data.

9.1 Introduction

Nonparametric methods offer an attractive alternative by allowing the data to "speak for themselves" without imposing rigid structural assumptions. These methods have gained considerable attention in recent decades due to their flexibility and ability to adapt to a wide variety of data patterns. In the context of time series analysis, nonparametric techniques can estimate unknown functions, such as conditional means or variances, directly from the observed data.

A particularly important class of time series processes is characterized by *long memory*, also known as long-range dependence or persistence. Long memory processes exhibit autocorrelations that decay slowly—typically at a hyperbolic rate rather than the exponential rate observed in short memory processes. This slow decay implies that observations far apart in time remain significantly correlated, a feature that has profound implications for statistical inference, forecasting, and economic interpretation.

Long memory has been documented in numerous empirical applications, including financial returns and volatility, inflation rates, interest rates, GDP growth, hydrological data, and network traffic. The presence of long memory fundamentally alters the statistical properties of estimators and test statistics, making it essential to account for this feature in the modeling process.

The combination of nonparametric methods with long memory time series presents both opportunities and challenges. On one hand, nonparametric

DOI: 10.1201/9781032627007-9

techniques can capture complex nonlinear dynamics that parametric long memory models might miss. On the other hand, the presence of long memory complicates the theoretical analysis of nonparametric estimators and requires modifications to standard bandwidth selection procedures.

This chapter provides an introduction to nonparametric methods for analyzing long memory time series. We begin in Section 9.2 by reviewing the general framework of nonparametric time series models, highlighting their advantages over parametric approaches. Section 9.3 formally introduces long memory processes and discusses their key properties. Section 9.4 presents kernel-based methods for nonparametric estimation in the time series context, while Section 9.5 describes commonly used kernel functions. The critical issue of bandwidth selection is addressed in Section 9.6, and Section 9.7 illustrates the role of the smoothing parameter through examples. Section 9.8 discusses applications to time series forecasting, and Section 9.9 presents simulation studies comparing nonparametric and parametric approaches. The chapter concludes with exercises in Section 9.10.

9.2 Nonparametric Time Series Models

Nonparametric methods in time series analysis aim to estimate unknown functions or relationships in the data without making strong assumptions about the underlying model structure. These methods provide a high degree of flexibility and can adapt to complex data patterns, which is particularly beneficial when the form of the time series process is not well understood. The flexibility of nonparametric models allows them to capture nonlinear dependencies, varying variances, and other irregular behaviors often present in real-world time series data. Unlike traditional parametric models, which assume a specific form for the data, nonparametric models do not require the imposition of specific functional forms, making them a powerful tool for modeling complex phenomena.

Nonparametric methods in time series analysis aim to estimate unknown functions or relationships in the data without making strong assumptions about the underlying model structure. These methods provide a high degree of flexibility and can adapt to complex data patterns, which is particularly beneficial when the form of the time series process is not well understood. The flexibility of nonparametric models allows them to capture nonlinear dependencies, varying variances, and other irregular behaviors often present in real-world time series data. Unlike traditional parametric models, which assume a specific form for the data, nonparametric models do not require the imposition of specific functional forms, making them a powerful tool for modeling complex phenomena.

Let X_t represent a time series dataset, where $t = 1, 2, 3, \ldots$ indexes time. The sequence of observations is given by:

$$X_t = (X_1, X_2, X_3, \ldots)$$

where each X_t is the observed value at time t. The time series data may exhibit patterns, trends, seasonality, or other dependencies over time.

In nonparametric time series, we model the relationship between the current observation X_t and its past values $X_{t-1}, X_{t-2}, \ldots, X_1$ using an unknown function $m()$, which can be either partially or entirely unknown. The general form of a nonparametric time series model can be expressed as:

$$X_t = m(X_{t-1}, X_{t-2}, \ldots, X_1) + \epsilon_t$$

where $m()$ is the unknown time series function and ϵ_t represents the error term. The function $m()$ could capture any nonlinearities or dependencies between the past and current values of the time series.

Example 1: First-order autoregression. Consider a first-order autoregressive process where the function $m()$ is entirely unknown. The equation can be written as:

$$X_t = m(X_{t-1}) + \epsilon_t$$

In this case, we assume that the current value of X_t depends on its previous value X_{t-1}, but the exact form of the function $m()$ is unknown and needs to be estimated nonparametrically. For example, this could capture nonlinear dependencies between X_t and X_{t-1} that a linear model would fail to represent.

Example 2: Second-order autoregression. For a second-order autoregressive process, the model could involve both X_{t-1} and X_{t-2}, with the function $m()$ still being unknown. The equation can be written as:

$$X_t = m(X_{t-1}, X_{t-2}) + \epsilon_t$$

Here, the current value X_t is a function of the two previous values X_{t-1} and X_{t-2}, but again, the form of the function $m()$ is unknown and estimated nonparametrically. This model allows for capturing more complex relationships and dependencies in the data without assuming a specific parametric structure.

In nonparametric time series analysis, the autocovariance structure of the time series is often unknown and must be estimated using the observed data. The autocovariance function, which describes the relationship between observations at different time points, plays a key role in characterizing the dependencies within the time series. Since the form of the autocovariance is typically unknown in nonparametric models, we rely on estimation techniques to approximate it from the available data. However, due to the complexity and intricacies involved in estimating the autocovariance, this aspect is not discussed in detail within this chapter. Readers interested in further exploration of this topic are encouraged to refer to advanced time series textbooks for more in-depth coverage of methods for estimating autocovariance and other related concepts.

One of the primary tools for nonparametric time series analysis is *kernel smoothing*, which involves estimating the underlying function of the data by averaging weighted local neighborhoods of the data points. Kernel methods have been widely applied to a variety of time series tasks, such as trend estimation, forecasting, and signal extraction. Fan and Yao (2003) provides an overview of nonparametric methods in time series analysis, detailing their flexibility and applications.

Fan and Gijbels (1996) intrduces kernel-based regression techniques that allow for the estimation of conditional expectations in time series data. These methods are particularly useful when the relationship between the variables is unknown or suspected to be nonlinear. Moreover, nonparametric models are advantageous when modeling heteroscedasticity, as they can capture nonconstant variance over time, which parametric models like GARCH may struggle to do effectively.

In addition to kernel smoothing, other nonparametric techniques have gained traction in time series analysis. For example, *wavelet transforms* are used to analyze time series data at different scales or frequencies, making them suitable for detecting patterns that vary over both time and frequency. Wavelets are particularly useful in capturing irregularities and short-term variations in data, which is a typical characteristic of financial or economic time series. Malfait and Roose (1997) provides an extensive review of the application of wavelets in time series analysis, particularly for nonstationary and fractal time series.

Spline methods, another important nonparametric technique, are often used for smoothing and interpolation. Splines are piecewise polynomial functions that can be fitted to data with minimal assumptions. Hastie and Tibshirani (1990) introduces spline smoothing in the context of regression and its application to time series data. Splines have proven to be effective in modeling data with abrupt changes or complex nonlinear trends, making them a useful tool in time series modeling.

With the advent of *machine learning*, more advanced techniques have also been applied to time series analysis. Methods such as *support vector machines (SVMs)* and *random forests* can model complex, high-dimensional time series data without relying on parametric assumptions. Machine learning models are particularly effective when dealing with large datasets or when the relationship between variables is highly complex and difficult to model parametrically. Bengio, et al. (2016) provides a comprehensive introduction to machine learning algorithms, many of which have applications in time series analysis.

The combination of these nonparametric techniques allows for greater flexibility and adaptability when modeling time series data, particularly when the data exhibit nonlinearities, long-range dependencies, or other complex behaviors. The key challenge, however, lies in the selection of appropriate methods and tuning parameters, such as bandwidth for kernel regression, the degree of smoothness for splines, or the number of layers in deep learning models.

9.3   Nonparametric Long Memory Time Series

Long memory processes, also known as persistent or long-range-dependent processes, are a class of time series models where the autocovariance function decays slowly, typically following a power-law decay rather than an exponential decay. This slow decay means that observations are not only influenced by their immediate neighbors but also by distant past values, leading to long-term dependencies that can persist over many periods. Long memory models are particularly useful for modeling processes that exhibit autocorrelation at large lags, such as financial, economic, and environmental time series data.

Beran (1994) first introduced the fractional autoregressive integrated moving average (FARIMA) model, a parametric model that generalizes the ARMA model by allowing for fractional differencing. The FARIMA model is commonly used to capture long memory in time series, as the fractional differencing parameter allows for a slow decay in the autocovariance function. However, while FARIMA models can capture long memory, they often assume linearity and stationarity, which may not always be appropriate for real-world data, especially in the presence of nonlinearities or structural changes.

Nonparametric methods have become increasingly important in the analysis of long memory processes. By avoiding strong assumptions about the structure of the data, nonparametric methods can provide a more flexible way of modeling long-range dependencies. One of the most effective nonparametric methods for capturing long memory in time series is kernel smoothing. Härdle (1992) investigates the use of kernel regression for modeling long memory time series and compares its performance to traditional parametric models. Kernel methods offer the advantage of being able to model complex, nonlinear long memory dependencies without requiring the specification of a parametric model.

In addition to kernel-based methods, other nonparametric approaches have been developed for modeling long memory time series. For example, Percival et al. (2000) presents a method based on wavelet transforms to capture long memory in nonstationary time series. The authors argue that wavelets are particularly effective in detecting long-range dependencies because they allow for multi-resolution analysis of time series data, making it easier to capture both short-term and long-term dependencies simultaneously.

Another nonparametric approach is the use of *fractional differencing* combined with kernel smoothing, which allows for the estimation of long memory processes while accounting for both nonlinearities and slow decaying autocovariances. Simpson and Taqqu (2006) explores the integration of fractional differencing with kernel smoothing techniques for more accurate modeling of long memory processes in financial data.

Machine learning techniques have also been applied to the analysis of long memory time series. Xu and Chen (2019) presents a machine learning-based approach to long memory modeling using deep learning architectures, such as recurrent neural networks (RNNs), which are well-suited for capturing temporal dependencies in time series data. These methods can be particularly useful when traditional parametric or nonparametric models are unable to capture the complex, nonlinear dynamics in long memory processes.

Overall, nonparametric methods offer a powerful and flexible framework for analyzing long memory time series. They provide the ability to model complex dependencies without imposing strict assumptions, making them particularly useful in real-world applications where the true underlying process is often unknown or difficult to model parametrically.

Zivot and Wang (2003) offers a comprehensive review of various approaches for modeling long memory and discusses the challenges and methodologies associated with analyzing such processes. The integration of nonparametric techniques, such as kernel methods, wavelets, and machine learning, with traditional long memory models provides a promising avenue for further research and application in the field of time series analysis.

9.4 Nonparametric Kernel-Based Time Series

Kernel-based methods are widely used in nonparametric time series analysis due to their flexibility and ability to model complex, nonlinear relationships between observations. In kernel-based approaches, the key idea is to estimate a function from the data by averaging local neighborhoods of data points, where the weights of the neighbors are determined by a kernel function. These methods do not impose parametric assumptions on the functional form of the data, making them ideal for situations where the underlying relationships are unknown or complex.

Kernel-based methods are widely used in nonparametric time series analysis due to their flexibility and ability to model complex, nonlinear relationships between observations. In kernel-based approaches, the key idea is to estimate a function from the data by averaging local neighborhoods of data points, where the weights of the neighbors are determined by a kernel function. These methods do not impose parametric assumptions on the functional form of the data, making them ideal for situations where the underlying relationships are unknown or complex.

For Example 1 (first-order autoregression), we aim to estimate the unknown function $m()$ using kernel-based methods. The first-order autoregressive model is expressed as:

$$X_t = m(X_{t-1}) + \epsilon_t$$

In kernel-based estimation, the function $m(X_{t-1})$ is estimated by averaging weighted local data points, where the weights are determined by a kernel function. A typical kernel function is a symmetric, non-negative function $K(u)$ that assigns higher weights to points closer to the target point and lower weights to points further away. The kernel estimator for $m(X_{t-1})$ at time t is given by:

$$\hat{m}(X_{t-1}) = \frac{\sum_{i=1}^{t-1} K\left(\frac{X_{t-1}-X_{i-1}}{h}\right) X_i}{\sum_{i=1}^{t-1} K\left(\frac{X_{t-1}-X_{i-1}}{h}\right)}$$

where:

- $K(u)$ is the kernel function, which is typically a smooth, symmetric function such as the Gaussian or Epanechnikov kernel. This kernel assigns weights to the neighboring points based on their distance from X_{t-1}.

- h is the bandwidth parameter, which controls the width of the neighborhood around X_{t-1} that is used for estimating $m(X_{t-1})$. A small value of h gives more weight to nearby points, while a larger h includes a broader range of data points, resulting in a smoother estimate.

- The summation runs over the previous observations up to $t-1$, as we are estimating $m(X_{t-1})$ using past data.

For Example 2 (second-order autoregression), the model involves two previous time points, X_{t-1} and X_{t-2}. The second-order autoregressive model is expressed as:

$$X_t = m(X_{t-1}, X_{t-2}) + \epsilon_t$$

In this case, kernel-based methods are applied to estimate the two-dimensional function $m(X_{t-1}, X_{t-2})$. The kernel estimator for $m(X_{t-1}, X_{t-2})$ at time t is

$$\hat{m}(X_{t-1}, X_{t-2}) = \frac{\sum_{i=1}^{t-1} K\left(\frac{X_{t-1}-X_{i-1}}{h_1}\right) K\left(\frac{X_{t-2}-X_{i-2}}{h_2}\right) X_i}{\sum_{i=1}^{t-1} K\left(\frac{X_{t-1}-X_{i-1}}{h_1}\right) K\left(\frac{X_{t-2}-X_{i-2}}{h_2}\right)}$$

where:

- $K(u)$ is the kernel function applied to both X_{t-1} and X_{t-2}.

- h_1 and h_2 are the bandwidth parameters for the first and second lags, respectively. These parameters control the smoothing for each lag. A small value for either bandwidth gives more weight to the most recent observations, while a larger value gives more weight to distant observations.

- The summation runs over the previous observations up to $t-1$, as we are estimating $m(X_{t-1}, X_{t-2})$ using past data.

Key insights

- The bandwidth parameters h, h_1, and h_2 control the smoothness of the estimators. A smaller bandwidth focuses the estimation on the most recent observations, leading to a less smooth estimate, while a larger bandwidth includes more distant points, resulting in a smoother estimate.

- The kernel function $K(u)$ determines the weight assigned to each observation based on its distance from the point being estimated. Common choices include the Gaussian and Epanechnikov kernels, which assign higher weights to nearby observations and lower weights to those farther away. We will introduce some common kernel functions in the next subsection.

The kernel-based approach provides a nonparametric way to estimate the unknown autoregressive function without assuming a specific parametric form for the relationship between the observations.

9.5 Kernel Functions

A kernel function is a symmetric, positive-definite function that assigns weights to data points based on their distance from a target point. The weight assigned to each data point decreases as the distance from the target point increases, with the rate of decrease determined by the kernel's bandwidth parameter. The kernel function, denoted by $K(x)$, is used to compute the weights in kernel-based estimators, such as kernel density estimation and kernel regression.

Several kernel functions have been proposed in the literature, and each one has different properties, making them more or less suited to different types of data. Below are some of the most commonly used kernel functions in nonparametric time series modeling:

- **Gaussian Kernel:** The Gaussian kernel is one of the most widely used kernels. It assigns exponentially decreasing weights as the distance from the target increases. The Gaussian kernel is defined as:

$$K_{\text{Gauss}}(x) = \frac{1}{\sqrt{2\pi}} \exp\left(-\frac{x^2}{2}\right)$$

- **Epanechnikov Kernel:** This kernel is known for its compact support, meaning that it assigns zero weight to points that are farther than a certain distance from the target. The Epanechnikov kernel is defined as:

$$K_{\text{Epanech}}(x) = \frac{3}{4}\left(1 - x^2\right)\mathbf{1}_{|x|\leq 1}$$

- **Uniform Kernel:** The uniform kernel assigns equal weights to all points within a fixed distance from the target and zero weight outside of that distance. The uniform kernel is defined as:

$$K_{\text{Uniform}}(x) = \frac{1}{2}\mathbf{1}_{|x|\leq 1}$$

- **Bi-weight Kernel:** Similar to the Epanechnikov kernel, the bi-weight kernel also has compact support and assigns more weight to the central points. The bi-weight kernel is defined as:

$$K_{\text{Biweight}}(x) = \frac{15}{16}\left(1 - x^2\right)^2 \mathbf{1}_{|x|\leq 1}$$

- **Tri-cube Kernel:** This kernel assigns a weight that decreases smoothly as the distance increases, with the rate of decrease being cubic in nature. The tri-cube kernel is defined as:

$$K_{\text{Tricube}}(x) = \frac{70}{81}\left(1 - |x|^3\right)^3 \mathbf{1}_{|x|\leq 1}$$

Figure 9.1 displays the plots of the five kernel functions mentioned above, illustrating their different shapes and how they assign weights to the data points.

9.6 Bandwidth Selection

In kernel-based methods, the bandwidth parameter plays a crucial role in determining the smoothness of the estimator. The bandwidth controls the width of the neighborhood around each point and, therefore, the extent to which neighboring points influence the estimation. A small bandwidth leads to a highly sensitive estimator that closely follows the data, while a large bandwidth results in a smoother estimator that may miss important local features.

Selecting an appropriate bandwidth is crucial for achieving good model performance. Bandwidth selection can be done using various methods, including cross-validation, plug-in methods, or using asymptotic formulas. Cross-validation methods often balance bias and variance by selecting a bandwidth that minimizes prediction error.

To select the bandwidth parameter h using cross-validation, the goal is to minimize the prediction error, balancing the bias and variance. The general approach involves the following steps:

Kernel Functions

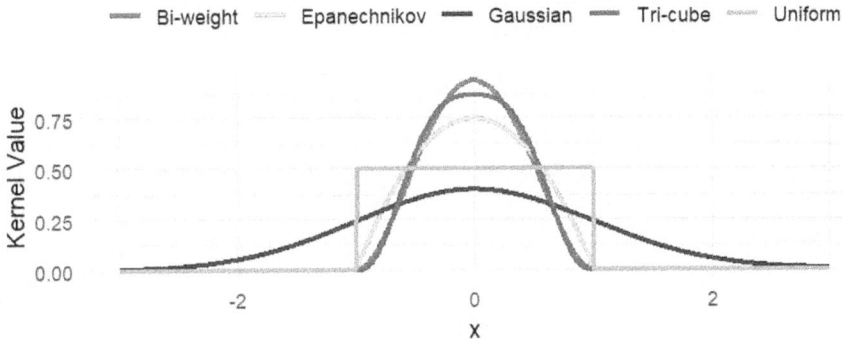

FIGURE 9.1

Plot of different kernel functions. Each kernel assigns weights to points based on their distance from the target point, with the bandwidth parameter controlling the width of the influence region. The Gaussian kernel (red) decays smoothly, while the Epanechnikov (green) and uniform (blue) kernels have compact support. The bi-weight (purple) and tri-cube (orange) kernels are also smooth but have varying rates of decay.

1. Split the time series data into a training set and a validation set.

2. For each candidate bandwidth h, estimate the kernel regression function on the training set.

3. Compute the prediction error on the validation set for each candidate bandwidth h.

4. Select the bandwidth that minimizes the prediction error.

The cross-validation error can be expressed as:

$$CV(h) = \frac{1}{T} \sum_{t=1}^{T} (X_t - \hat{m}(X_{t-1}, X_{t-2}, \dots))^2$$

where:

- T is the number of data points (excluding the first few that cannot be predicted due to lack of prior observations).

- X_t is the observed value of the time series at time t.

- $\hat{m}(X_{t-1}, X_{t-2}, \dots)$ is the kernel estimator of the function, calculated using the training set with the candidate bandwidth h.

- The sum is over all the time points in the validation set.

The optimal bandwidth h_{opt} is the value of h that minimizes the cross-validation error:

$$h_{opt} = \arg\min_h CV(h)$$

This formula finds the bandwidth that results in the smallest average prediction error over the validation set, providing a good trade-off between bias and variance.

In practice, you would evaluate the cross-validation error for different values of h and choose the one that gives the minimum error, ensuring that the kernel-based model is well-tuned for the time series data.

9.6.1   R code implementation

The following R code is used to implement the kernel-based estimator and perform bandwidth selection via cross-validation:

```r
# Load necessary libraries
library(stats)

# Simulate some time series data (AR(1) process)
set.seed(123)
n <- 100  # number of observations
phi <- 0.8  # autoregressive parameter
epsilon <- rnorm(n)  # random noise
X <- numeric(n)

# Generate the time series using AR(1) process
for (t in 2:n) {
  X[t] <- phi * X[t-1] + epsilon[t]
}

# Kernel estimator function
kernel_estimator <- function(X, h) {
  # Calculate kernel weights based on the bandwidth h (Gaussian kernel)
  kernel_weights <- function(x, h) {
    exp(-0.5 * (x / h)^2) / (h * sqrt(2 * pi))
  }

  # Estimate function m(X_t-1) for each X_t using kernel weights
  m_hat <- numeric(length(X))
  for (t in 2:length(X)) {
    weights <- kernel_weights(X[t-1] - X[1:(t-1)], h)
    m_hat[t] <- sum(weights * X[1:(t-1)]) / sum(weights)
  }
  return(m_hat)
}

# Cross-validation function for bandwidth h
cross_validation <- function(h, X) {
  m_hat <- kernel_estimator(X, h)
```

Cross-validation for Bandwidth Selection

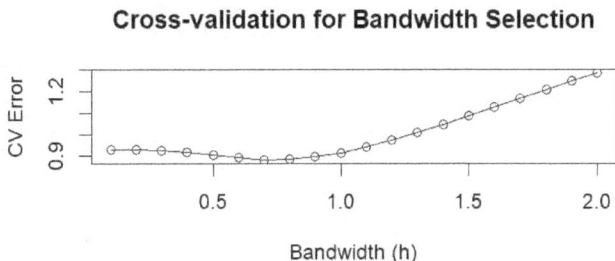

FIGURE 9.2
Cross-validation error for bandwidth selection.

```
cv_error <- mean((X[2:length(X)] - m_hat[2:length(m_hat)])^2)
return(cv_error)
}

# Range of bandwidth values to test
h_values <- seq(0.1, 2, by = 0.1)

# Minimize cross-validation error by finding optimal h
cv_errors <- sapply(h_values, function(h) cross_validation(h, X))

# Find optimal bandwidth
h_opt <- h_values[which.min(cv_errors)]
cat("Optimal bandwidth h:", h_opt, "\n")

# Plot the cross-validation error vs. bandwidth
plot(h_values, cv_errors, type = "o", col = "blue", xlab = "Bandwidth (h)",
 ylab = "CV Error", main = "Cross-validation for Bandwidth Selection")
```

The optimal bandwidth h that minimizes the cross-validation error is obtained by applying the cross-validation procedure described above. The plot below (Figure 9.2) illustrates the cross-validation error as a function of bandwidth values, showing the bandwidth value that results in the smallest error.

The optimal bandwidth value, h_{opt}, is selected based on this plot, providing the best balance between bias and variance in the kernel regression estimator.

9.7 The Role of the Smooth Parameter in Estimation

To illustrate how the bandwidth affects the estimation, let us consider a simple example where we generate a synthetic time series from a known function and estimate it using kernel smoothing. By adjusting the bandwidth, we can observe the effect of smoothing on the kernel estimator.

Example:

This example illustrates the importance of selecting an appropriate bandwidth for nonparametric time series estimation. The choice of bandwidth determines the balance between capturing the underlying pattern of the data and avoiding overfitting to noise.

```
# Load required libraries
library(ggplot2)
library(np) # For nonparametric kernel regression

# Set seed for reproducibility
set.seed(123)

# Generate synthetic AR(1) time series
n <- 200 # Number of observations
X <- numeric(n)
X[1] <- rnorm(1) # Initial value
phi <- 0.8 # AR(1) coefficient
for (t in 2:n) {
  X[t] <- phi * X[t-1] + rnorm(1, sd = 0.5) # AR(1) process with noise
}

# Create a data frame for the time series
data <- data.frame(
  Time = 1:n,
  X = X
)

# Kernel smoothing function
kernel_smooth <- function(X, bandwidth) {
  # Use npreg for kernel regression
  model <- npreg(X ~ Time, data = data, bws = bandwidth, ckertype =
  "gaussian")
  return(fitted(model))
}

# Apply kernel smoothing with different bandwidths
bandwidths <- c(0.1, 1, 10) # Different bandwidths
smoothed_data <- lapply(bandwidths, function(bw) kernel_smooth(X, bw))

# Add smoothed data to the data frame
data$Smoothed_0.1 <- smoothed_data[[1]]
data$Smoothed_1 <- smoothed_data[[2]]
data$Smoothed_10 <- smoothed_data[[3]]

# Plot the original time series and smoothed estimates
ggplot(data, aes(x = Time)) +
  geom_line(aes(y = X, color = "Original Data"), size = 0.8) +
  geom_line(aes(y = Smoothed_0.1, color = "Bandwidth = 0.1"), size = 1) +
```

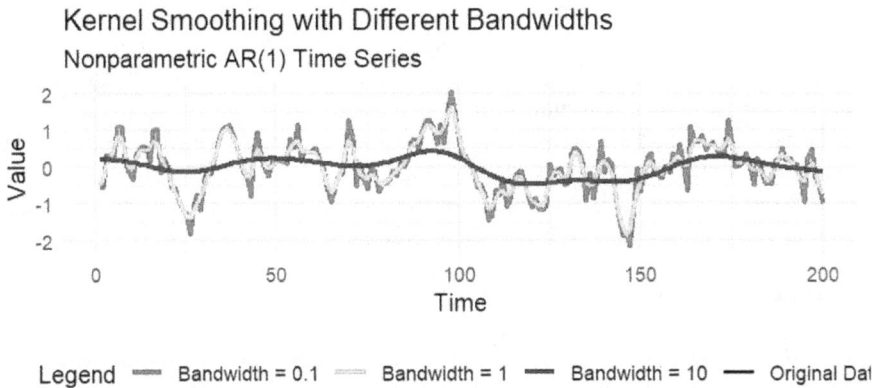

FIGURE 9.3
Kernel smoothing with different bandwidths.

```
geom_line(aes(y = Smoothed_1, color = "Bandwidth = 1"), size = 1) +
geom_line(aes(y = Smoothed_10, color = "Bandwidth = 10"), size = 1) +
labs(
    title = "Kernel Smoothing with Different Bandwidths",
    subtitle = "Nonparametric AR(1) Time Series",
    x = "Time",
    y = "Value",
    color = "Legend"
) +
scale_color_manual(values = c(
    "Original Data" = "black",
    "Bandwidth = 0.1" = "red",
    "Bandwidth = 1" = "green",
    "Bandwidth = 10" = "blue"
)) +
theme_minimal() +
theme(legend.position = "bottom")
```

Figure 9.3 shows the plot generated from the above R code. It compares the original noisy time series and the smoothed estimates using different bandwidth values. The different colors correspond to different bandwidths, with the black line representing the true underlying function.

9.8   Application to Time Series Forecasting

Kernel-based methods can be applied to time series forecasting by estimating the underlying time series function or model.

One important aspect of kernel-based time series forecasting is the selection of an appropriate kernel function and bandwidth. Different kernels and bandwidths will lead to different forecast behaviors, making it crucial to test and choose the best combination for the specific data and problem at hand.

In the following sections, we will explore examples that illustrate how kernel-based time series methods work.

9.9 Simulation Examples

In this section, we present two more examples of kernel-based time series estimation and forecasting with long memory.

Example 4: Long-Memory Time Series Generation and Forecasting

We first generate a long-memory time series using the FARIMA model and estimate the time series function using kernel regression. The following steps were taken:

1. Generate a long-memory time series using fractional differencing. 2. Estimate the time series function using kernel regression. 3. Forecast future values based on the kernel model.

Example 5: Comparison of Kernel-Based Estimators with Parametric Models

We compare the performance of kernel-based nonparametric estimators with conventional parametric time series models, specifically using ARFIMA and FARIMA models. This comparison highlights the advantages of kernel methods in capturing complex nonlinear relationships and long memory dependencies.

9.10 Chapter 9 Questions

Question 1: Simulating a Nonlinear Long Memory Time Series and Estimating with Kernel-Based Method

Generate a nonlinear long-memory time series where X_t is a function of its past values. For example, simulate the process with the following nonlinear function:

$$X_t = \sin(0.5X_{t-1}) + 0.3X_{t-2} + \epsilon_t,$$

where ϵ_t is white noise. Fit a nonparametric kernel-based estimator to this time series and evaluate the smoothed function. Then, forecast the next 10 time points using the kernel estimator.

Question 2: Comparing Kernel Estimator and ARFIMA Model for Long Memory Time Series

Using the time series generated in Question 1, fit both a kernel-based estimator and an ARFIMA model. Compare their performance in forecasting the next 10 time points. Discuss the differences between the two approaches and the accuracy of their forecasts.

Question 3: Simulation of Long Memory Time Series with ARFIMA and Kernel-Based Estimators

Generate a long-memory time series using the ARFIMA(0, d, 0) model where $d = 0.4$. Estimate the function using a kernel-based nonparametric estimator and compare the results with the ARFIMA model. Make 10-step ahead forecasts using both models and evaluate the forecast accuracy.

Question 4: Forecasting with Kernel and ARFIMA Models on Real Data (Airline Passenger Data)

Download the airline passenger dataset available at `https://raw.githubusercontent.com/jbrownlee/Datasets/master/airline-passengers.csv`. This dataset consists of monthly totals of international airline passengers from 1949 to 1960. Fit both a kernel-based nonparametric estimator and an ARFIMA model to the data. Forecast the next 12 months and compare the two methods in terms of forecast accuracy.

Question 5: Forecasting with Kernel and ARFIMA Models on Real Data (US Macro-Economic Data)

Download the US macroeconomic data (e.g., GDP, inflation) from `https://fred.stlouisfed.org/`. Use any economic indicator of your choice and apply both kernel-based nonparametric time series models and ARFIMA models to predict the next quarter/year. Compare the performance of both models and discuss the results.

10

ARMA, ARIMA, ARFIMA, and GARMA Models with GARCH Errors

Synopsis: An optimal estimation methodology based on state space modeling of long memory Gegenbauer processes driven by conditionally heteroskedastic errors is seemingly absent in the current literature. In lieu of it, this chapter considers an approximation of long memory Gegenbauer processes driven by heteroskedastic errors using finite-order moving-average (MA) and autoregressive (AR) representations. A related state space form is used to estimate parameters by pseudo maximum likelihood and the Kalman filter. As a novel contribution, a comparative assessment of the suggested approximation techniques is performed using a large scale simulation study. It results in extensive Monte Carlo evidence that establish and validate the optimal order of the two approximations as interval estimates. The superiority of the created methodology is illustrated as an original contribution by extending it to nonstationary Gegenbauer processes to compare with similar existing estimation techniques and results in the literature. Finally, the approach is applied to the famous daily Standard and Poor (S and P) 500 series as a real application.

10.1 Notation and Preliminaries

This section establishes the notation and reviews the preliminary concepts used throughout this chapter. Let $\{X_t\}_{t=1}^n$ denote a time series of n observations. We use B to denote the backshift operator, defined by $BX_t = X_{t-1}$, and more generally $B^k X_t = X_{t-k}$ for any integer k. The difference operator is denoted $\Delta = 1 - B$, so that $\Delta X_t = X_t - X_{t-1}$.

For a sequence of random variables $\{Y_t\}$, we write $Y_t \sim \text{NID}(0, 1)$ to indicate that the Y_t are independently and identically distributed with mean zero and variance one. The notation $Y_t | \mathcal{F}_{t-1} \sim N(0, h_t)$ indicates that Y_t, conditional on the information set $\mathcal{F}_{t-1}$, follows a normal distribution with mean zero and variance h_t. We use $\mathbb{Z}^+$ to denote the set of positive integers, and $[x]$ denotes the integer part of a real number x.

DOI: 10.1201/9781032627007-10

The ARFIMA(p, d, q) model generalizes the standard ARIMA framework by allowing the differencing parameter d to take non-integer values in the interval $(-0.5, 0.5)$. The fractional differencing operator $(1 - B)^d$ is defined through its binomial expansion:

$$(1 - B)^d = \sum_{k=0}^{\infty} \frac{\Gamma(k - d)}{\Gamma(-d)\Gamma(k + 1)} B^k,$$

where $\Gamma(\cdot)$ denotes the gamma function. The Gegenbauer ARMA model, GARMA(p, d, q), further extends this framework by replacing the standard differencing operator with the Gegenbauer polynomial $(1 - 2uB + B^2)^d$, where $|u| < 1$ determines the frequency at which long memory occurs.

The GARCH(r, s) process for the error term ε_t is defined as $\varepsilon_t = \epsilon_t \sqrt{h_t}$, where $\epsilon_t \sim NID(0, 1)$ and the conditional variance satisfies $h_t = \alpha_0 + \sum_{i=1}^{r} \alpha_i \varepsilon_{t-i}^2 + \sum_{j=1}^{s} \beta_j h_{t-j}$. For covariance stationarity, we require $\alpha_0 > 0$, $\alpha_i \geq 0$, $\beta_j \geq 0$, and $\sum_{i=1}^{r} \alpha_i + \sum_{j=1}^{s} \beta_j < 1$.

A key tool in our analysis is the state space representation, consisting of an observation equation $X_t = Z\alpha_t + \varepsilon_t$ and a transition equation $\alpha_{t+1} = T\alpha_t + H\eta_t$, where α_t is the unobserved state vector and Z, T, H are system matrices. The Kalman filter provides an efficient recursive algorithm for computing the likelihood function of models in this form, which is central to the quasi-maximum likelihood estimation methodology developed in Section 10.5.

10.2 The Stationary Case

Consider a GARMA$(0, d, 0)$-GARCH(r, s) process defined as:

$$(1 - 2uB + B^2)^d X_t = \varepsilon_t, \tag{10.1}$$

where $|u| < 1$ and $0 < d < 1/2$, $\varepsilon_t | F_{t-1} \sim N(0, h_t)$ (F_{t-1} is the history of the process) and the conditional variance h_t satisfies the GARCH(r, s) process such that

$$\varepsilon_t = \epsilon_t \sqrt{h_t}, \quad \epsilon_t \sim NID(0, 1), \tag{10.2}$$

and

$$h_t = \alpha_0 + \sum_{i=1}^{r} \alpha_i \varepsilon_{t-i}^2 + \sum_{j=1}^{s} \beta_j h_{t-j}, \tag{10.3}$$

where $\alpha_0 > 0$, all $\alpha_j \geq 0$, all $\beta_j \geq 0$.

When $u = 1$ in equation (10.1), the process becomes an ARFIMA(0, d_*, 0)-GARCH(r,s) process defined as:

$$(1 - B)^{2d} X_t = \varepsilon_t = (1 - B)^{d_*} X_t = \varepsilon_t, \tag{10.4}$$

where $d_* = 2d$.

If d_* in (10.4) becomes an integer differencing operator instead of a fractional differencing operator then the ARFIMA(0, d_*, 0)-GARCH(r,s) process reduces to an ARIMA(0, d_*, 0)-GARCH(r,s) process. Thereafter, if the differencing filter $(1 - B)^{d_*}$ is removed from the process it reduces to a short memory ARMA(0, 0)-GARCH(r,s) process.

The theoretical (true) acf of the equation (10.1) could be found as a closed-form solution utilizing Gegenbauer coefficients through the expression

$$E(X_k X_{k+t}) = \sum_{j=0}^{\infty} \sum_{i=0}^{\infty} C_j C_i [E(\varepsilon_{t-j})(\varepsilon_{k+t-i})]. \tag{10.5}$$

Need to run the following code in R set.seed(1)

```
require(TSA)
t=numeric(400)
x=garch.sim(alpha=c(0.4,0.3),beta=c(0.3),n=1000)
y=x[201:1000]
p=numeric(400)
coeff=numeric(800)
g=numeric(400)
for(j in 3:800)
l=0.45
v=0.8
coeff[1]=1
coeff[2]=0.72
coeff[j]=(((2*v)*(j+l-1)*coeff[j-1])-((j+(2*l)-2)*coeff[j-2]))/j

cone=coeff[1:400]
ctwo=coeff[401:800]
k=1:400
for(i in 401:800)
g[i]=sum(cone[k]*y[i-k])
t=g[401:800]

for(n in 1:400)
p[n]=sum(cone[k]*ctwo[n])

Realization=ts.plot(t)olatility
par(mfrow=c(2,2))
ts.plot(t)
```

Series t

Time

Lag

Series p   **Series t**

Lag   Lag

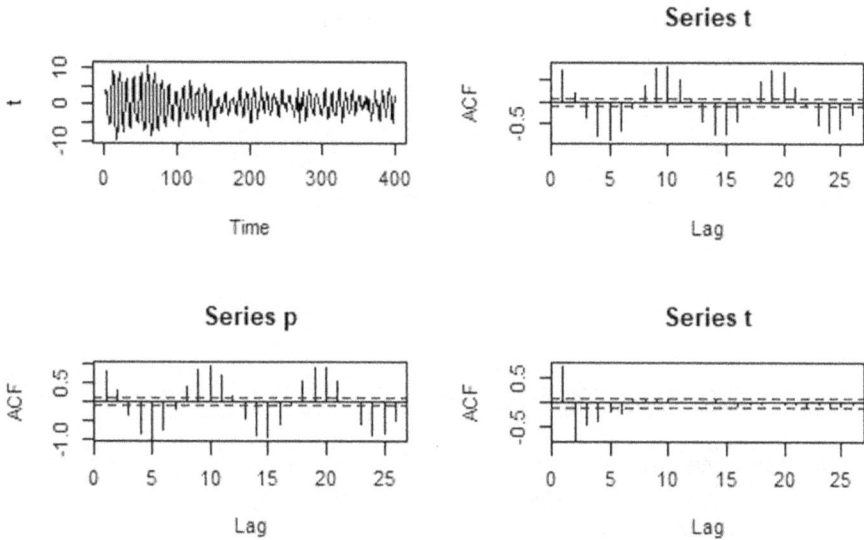

FIGURE 10.1
Images of Realization, sample acf, true acf, and pacf of a GARMA(0,1,0)-GARCH(1,1) Series.

```
acf(t)
acf(p)
acf(t,type="partial")
dev.off()
```

Running the above R code would generate simulation graphic grids depicting 400 length realizations (top left), sample acf (top right), true acf (bottom left), and pacf (bottom right) for examples of the introduced class with differing sets of statistical parameters as shown in Figure 10.1. They were developed in R by extending the ideas of Shumway and Stoffer (2010) to provide a visual illustration of properties such as volatility, long memory, and conditional heteroskedasticity.

By inspecting Figure 10.1, it is evident that these plots show a very high degree of volatility and conditional heteroskedasticity. By inspecting the sample and theoretical autocorrelations, it is clear that they are similar. Furthermore, it is clear that all the partial autocorrelation functions above decay slowly at a hyperbolic rate characterizing long memory.

The following theorem provides some basic properties of the process introduced in equations (10.1)–(10.3):

Theorem 10.1. Suppose $\alpha_0 > 0$, $\alpha_i \geq 0$, $\beta_i \geq 0$, $\sum_{i=1}^{r} \alpha_i + \sum_{j=1}^{s} \beta_j < 1$, $|u| = 1$ and $d < 1/4$. Then, for the process defined in (10.1)–(10.3), there exists a F_t-measurable second-order stationary solution $\{\varepsilon_t, X_t\}$ and it is the only second-order stationary solution given the ϵ_ts. The solution $\{\varepsilon_t, X_t\}$ has the following causal representations:

$$\varepsilon_t = \epsilon_t \{\alpha_0 + \sum_{j=1}^{\infty} \delta^T (\prod_{i=1}^{j} A_{t-i}) \mathcal{E}_{t-j}\}^{1/2} \quad a.s., \tag{10.6}$$

and

$$X_t = \sum_{j=0}^{[M/2]} C_j \varepsilon_{t-j} \quad a.s., \tag{10.7}$$

where

- $\mathcal{E}_t = (\alpha_0 \epsilon_t^2, 0, ...0, \alpha_0, 0, ..., 0)_{(r+s) \times 1}^T$,

- the first component is $\alpha_0 \epsilon_t^2$ and the (r+1)th component is α_0, $\{\epsilon_t\}$ are independently normally distributed with mean 0 and variance 1, $\delta = (\alpha_1, ..., \alpha_r, \beta_1, ..., \beta_s)^T$, and

$$A_t = \begin{bmatrix} \alpha_1 \epsilon_t^2 & \cdots & \alpha_r \epsilon_t^2 & \beta_1 \epsilon_t^2 & \cdots & \beta_s \epsilon_t^2 \\ I_{(r-1) \times (r-1)} & O_{(r-1) \times 1} & & O_{(r-1) \times s} & \\ \alpha_1 & \cdots & \alpha_r & \beta_1 & \cdots & \beta_s \\ O_{(s-1) \times r} & & I_{(s-1) \times (s-1)} & O_{(s-1) \times 1} \end{bmatrix},$$

where • $\mathbf{I_{r \times r}}$ is the $r \times r$ identity matrix.

Proof. For the proof of Theorem 10.1, reference could be made to the proof of Theorem 2.1 given in the appendix of Ling and Li (1997) to a similar specific member that is applicable to the generalized class presented in this chapter.   □

Lemma 10.1. $\{\varepsilon_t\}$ and $\{X_t\}$ in (1) are stationary.

The next theorem shows some additional properties of the GARMA$(0, d, 0)$-GARCH(r,s) model introduced in (10.1)–(10.3).

Theorem 10.2. Let $\{X_t\}$ be generated by (10.1)–(10.3). Assume roots of associated autoregressive and moving-average operators lie outside the unit circle, and $\sum_{i=1}^{r} \alpha_i + \sum_{j=1}^{s} \beta_j < 1$:

(a.) If $d < 1/2$, then $\{X_t\}$ is second-order stationary with the generic MA approximation representation:

$$X_t = \sum_{j=0}^{[M/2]} C_j \varepsilon_{t-j}, \tag{10.8}$$

where ε_t has representation (10.6). Hence, $\{X_t\}$ is stationary.

(b.) If $d > -1/2$, then $\{X_t\}$ is invertible; that is, ε_t can be written as the generic AR approximation representation:

$$\varepsilon_t = \sum_{j=0}^{[M/2]} \pi_j X_{t-j}, \qquad (10.9)$$

where $\pi(B) = (1 - 2uB + B^2)^d$.

Proof. For part (a.) Let $\eta_1(Z) = (1 - 2uZ + Z^2)^{-d}$. The closed-form solution of the recursive definition of $(1 - 2uZ + Z^2)^{-d}$ converge for $|Z| \leq 1$. Hence, $\{X_t\}$ exists with representation (10.7). Refer Theorem (10.1) and proof of Theorem (2.2) of Ling and Li (1997) that establishes $\{\varepsilon_t\}$ is second-order stationary. By utilizing the same results of Ling and Li (1997), it could be shown that $\{X_t\}$ is also second-order stationary. From representation (6), ε_t is a measurable function of iid random variable Z_ts and hence so is $\{X_t\}$. Therefore, $\{X_t\}$ is stationary. For part (b.) Let $\eta_2(Z) = (1 - 2uZ + Z^2)^d$. Similar to η_1, the function η_2 converges, when $|Z| \leq 1$ and (10.8) holds true. □

Note: On certain occasions, the nonstationary version of the model (10.1) becomes useful from a meta-analytical perspective and becomes the next discussion topic.

10.3   The Nonstationary Case

The model in (10.1) could be reduced to a nonstationary version of a GARMA$(0, d, 0)$-GARCH(r,s) series defined by

$$(1 - 2uB + B)^{d_1}(1 - 2uB + B)^m X_t = \varepsilon_t, \qquad (10.10)$$

where $|u| < 1$, $|d_1| < 1/2$, $m \in Z^+$, ε_t represents GARCH errors.

10.4   MA and AR Approximations

The Wold representation of the GARMA$(0, d, 0)$-GARCH(r,s) process in (10.1) is

$$X_t = \sum_{j=0}^{\infty} C_j \varepsilon_{t-j}, \qquad (10.11)$$

The m-th-order moving-average approximation of the GARMA$(0, d, 0)$-GARCH(r,s) process is

$$X_{t,m} = \sum_{j=0}^{m} C_j \varepsilon_{t-j}. \qquad (10.12)$$

$X_{t,m}$ will be referred to as a truncated Gegenbauer process; the coefficients C_j are functionally dependent on d and u. For simplicity, we replace $X_{t,m}$ by X_t hereafter.

The corresponding state space representation of the MA(m) model is

$$\begin{aligned} X_t &= Z\alpha_t + \varepsilon_t, \\ \alpha_{t+1} &= T\alpha_t + H\varepsilon_t, \end{aligned} \qquad (10.13)$$

where the system matrices are

$$Z = [1, 0, \ldots, 0], \quad T = \begin{bmatrix} 0 & 1 & 0 & \cdots & 0 \\ 0 & 0 & 1 & \ddots & 0 \\ \vdots & \vdots & \ddots & \ddots & 0 \\ \vdots & \cdots & \cdots & 0 & 1 \\ 0 & 0 & \cdots & \cdots & 0 \end{bmatrix}, \quad H = \begin{bmatrix} C_1 \\ C_2 \\ \vdots \\ \vdots \\ C_m \end{bmatrix}$$

for suitably chosen matrices Z, α_t, T and H with dimensions $1 \times m, m \times 1, m \times m$, and $m \times 1$. The vector $\alpha_t = [X(t|t-1), X(t+1|t-1), X(t+2|t-1), \ldots, X(t+m-1|t-1)]'$ consists of observations conditioned upon state at $t-1$. These system matrices typically depend on hyperparameters. The first and second equations of (10.13) are the *Measurement (or Observation)* and *Transition (or State)* equations of the process.

We may also derive the AR(m) approximation by truncating the AR(∞) representation $\pi(B)X_t = \varepsilon_t, \pi(B) = (1 - 2uB + B^2)^d$. We compare these two approximations with Chan and Palma (1998) and Grassi and De Magistris (2014) for the autoregressive fractionally integrated moving average (ARFIMA) case when $u = 1$. The relevant estimation methodology is provided in the next section.

10.5   Quasi-Maximum-Likelihood Estimation Methodology

Given a sample time series $\{x_t, t = 1, \ldots, n\}$, the likelihood function of the approximating MA(m) model represented in (10.11) is evaluated with the support of the KF, which is the following set of recursions ($t = 1, \ldots, n$),

$$\begin{aligned} \nu_t &= x_t - Za_t, & f_t &= ZP_tZ', \\ & & K_t &= (TP_tZ')/f_t, \\ a_{t+1} &= Ta_t + K_t\nu_t, & P_{t+1} &= TP_tT' + HH' - K_tK_t'/f_t. \end{aligned} \qquad (10.14)$$

The KF returns the pseudo-innovations ν_t, such that if the MA(m) approximation were the true model, $\nu_t \sim \mathrm{NID}(0, \varepsilon f_t)$, so that the log-likelihood of (d, u, ε) is (apart from a constant term)

$$\ell(d, u, \varepsilon) = -\frac{1}{2}\left(n \ln \varepsilon + \sum_{t=1}^{n} \ln f_t + \frac{1}{\varepsilon}\sum_{t=1}^{n}\frac{\nu_t^2}{f_t}\right). \tag{10.15}$$

The scale parameter ε can be concentrated out of the likelihood function, so that

$$\hat{\varepsilon} = \sum_t \frac{\nu_t^2}{f_t}, \tag{10.16}$$

and the profile likelihood is

$$\ell_\varepsilon(d, u) = -\frac{1}{2}\left[n(\ln \hat{\varepsilon} + 1) + \sum_{t=1}^{n}\ln f_t\right]. \tag{10.17}$$

Chan and Palma (1998) states MA and AR approximations within a state space configuration delivers QMLE estimates of parameters due to an approximate likelihood function. Restricting parameters α_0, α_1, and β within a state space configuration becomes a cumbersome exercise and affects overall accuracy. Therefore, a built-in GARCH error fitting function known as *"garchfit"* in MATLAB, which returns QMLE estimates due to the conditional variance of a GARCH model as described by Ho and Houmani (2010), was incorporated with the state space modeling configuration. Monte Carlo evidence based on the introduced model for simulated data is given in the next section to highlight optimal and positive attributes of the considered time series.

10.6 Monte Carlo Evidence

An extensive Monte Carlo experiment on simulated data was conducted to estimate the parameters d, u, α_0, α_1, and β. A GARMA$(0, d, 0)$-GARCH$(1,1)$ process was considered in all the experiments for practical convenience. The experiment was later extended to establish the optimal estimation order of the process, distinguish the better approximation technique, and make a comparative assessment with other available results of similar methods in the literature.

In presenting the experiment results, let v be the number of iterations (in the tables below as well), n the series length and θ a generic term for d, u, α_0, α_1, and β. The following computations were carried out from the simulation study. Approximations of each measure are depicted as a subscript or included in brackets as "MA" or "AR" hereafter in all the tables.

(1) *Mean* $\hat{\theta} = \frac{1}{v}\sum_{i=1}^{v}\theta$.

(2) *Standard Error (SE)* $= (\frac{1}{v-1}\sum_{i=1}^{v}(\theta - \hat{\theta})^2)^{1/2}$.

(3) *Mean Square Error MSE* $=\frac{1}{v}\sum_{i=1}^{v}(\hat{\theta} - \theta)^2$.

Note: The benchmark for the model MSE was the total trace parameter MSE (TTP-MSE) defined as the sum of the MSEs of all trace parameters of an estimator variance–covariance matrix. It is denoted as:

(4) $TTP - MSE = MSE(d) + MSE(u) + MSE(\alpha_0) + MSE(\alpha_1) + MSE(\beta)$.

(5) $RMSE = \sqrt{MSE}$.

Note: The benchmark for the model RMSE was the total trace parameter RMSE (TTP-RMSE) defined as the sum of the RMSE's of all trace parameters of an estimator variance–covariance matrix. It is denoted as:

(6) $TTP - RMSE = RMSE(d) + RMSE(\alpha_0) + RMSE(\alpha_1) + RMSE(\beta)$ [Used in the analysis below of the Special Case (*) ARFIMA(0,d_*,0)-GARCH(r,s) process, when $u = 1$ in model of (1) with $d_* = 2d$].

The estimates of Tables 10.1–10.3 were arrived at by taking into consideration the TTP-MSE and the validating maximum likelihood value (*Likelihood*) as depicted by confirmatory examples in Table 10.4. The *Likelihood* is an output of the KF, while the TTP-MSE is a sum of mse values with respect to parameters d, u, α_0, α_1, and β. Therefore, it makes the validation process extremely plausible and feasible. In Table 10.4, if the TTP-MSE values are equal up to three decimal places, then the value corresponding to the lag order (m) in the middle will have the smallest magnitude up to four decimal places. The result of Table 10.4 is used to establish the optimal lag orders given in Table 10.5.

Note: Another interesting result from the MA approximation results given above is that the asymptotic variance of the long memory parameter (d) estimates of any stationary Gegenbauer series is approximately equal to $\frac{\pi^2}{12n}$ for any value of n.

Remark 10.1. Tables 10.6–10.8 are based on a nonstationary ARFIMA $(0, d_*, 0) - GARCH(r, s)$ model. It is a special case of a GARMA(0, d, 0)-GARCH(r, s) model and was taken into consideration, since estimation results of the considered exact generalized model is missing in the literature. The exercise was adequate for a comparative meta-analysis assessment of the newly introduced state space approximation techniques of this chapter.

Note: Results of Tables 10.1–10.5 and implementation of Wold expansion and KF recursions show that the MA approximation technique is better than it's AR counterpart in terms of ease of implementation and optimal lag order. Additionally processing speeds of 9 and 11 minutes approximately for MA

TABLE 10.1
QMLE estimates for stationary case with d = 0.1, u = 0.8, $\alpha_0 = 0.4$, $\alpha_1 = 0.3$, $\beta = 0.3$, and $v = 1000$

				$n = 100$						
Estimator	$\hat{d}_{MA}$	$\hat{u}_{MA}$	$\hat{\alpha}_{0MA}$	$\hat{\alpha}_{1MA}$	$\hat{\beta}_{MA}$	$\hat{d}_{AR}$	$\hat{u}_{AR}$	$\hat{\alpha}_{0AR}$	$\hat{\alpha}_{1AR}$	$\hat{\beta}_{AR}$
Mean	0.146	0.520	0.423	0.277	0.297	0.141	0.527	0.428	0.272	0.293
SE	0.086	0.497	0.149	0.182	0.097	0.080	0.490	0.152	0.181	0.096
MSE	0.009	0.325	0.022	0.033	0.009	0.008	0.314	0.024	0.033	0.009
				$n = 200$						
Estimator	$\hat{d}_{MA}$	$\hat{u}_{MA}$	$\hat{\alpha}_{0MA}$	$\hat{\alpha}_{1MA}$	$\hat{\beta}_{MA}$	$\hat{d}_{AR}$	$\hat{u}_{AR}$	$\hat{\alpha}_{0AR}$	$\hat{\alpha}_{1AR}$	$\hat{\beta}_{AR}$
Mean	0.119	0.659	0.422	0.276	0.297	0.122	0.643	0.416	0.282	0.296
SE	0.058	0.358	0.136	0.165	0.090	0.060	0.364	0.133	0.162	0.083
MSE	0.003	0.148	0.019	0.027	0.008	0.004	0.157	0.018	0.026	0.007
				$n = 500$						
Estimator	$\hat{d}_{MA}$	$\hat{u}_{MA}$	$\hat{\alpha}_{0MA}$	$\hat{\alpha}_{1MA}$	$\hat{\beta}_{MA}$	$\hat{d}_{AR}$	$\hat{u}_{AR}$	$\hat{\alpha}_{0AR}$	$\hat{\alpha}_{1AR}$	$\hat{\beta}_{AR}$
Mean	0.107	0.757	0.415	0.281	0.299	0.109	0.756	0.417	0.283	0.297
SE	0.039	0.213	0.103	0.123	0.063	0.038	0.200	0.103	0.123	0.064
MSE	0.001	0.047	0.010	0.015	0.004	0.001	0.042	0.011	0.015	0.004
				$n = 1000$						
Estimator	$\hat{d}_{MA}$	$\hat{u}_{MA}$	$\hat{\alpha}_{0MA}$	$\hat{\alpha}_{1MA}$	$\hat{\beta}_{MA}$	$\hat{d}_{AR}$	$\hat{u}_{AR}$	$\hat{\alpha}_{0AR}$	$\hat{\alpha}_{1AR}$	$\hat{\beta}_{AR}$
Mean	0.102	0.790	0.406	0.292	0.301	0.102	0.784	0.408	0.290	0.301
SE	0.028	0.117	0.067	0.083	0.045	0.028	0.127	0.072	0.090	0.045
MSE	0.0008	0.013	0.004	0.007	0.002	0.0007	0.016	0.005	0.008	0.002
				$n = 2000$						
Estimator	$\hat{d}_{MA}$	$\hat{u}_{MA}$	$\hat{\alpha}_{0MA}$	$\hat{\alpha}_{1MA}$	$\hat{\beta}_{MA}$	$\hat{d}_{AR}$	$\hat{u}_{AR}$	$\hat{\alpha}_{0AR}$	$\hat{\alpha}_{1AR}$	$\hat{\beta}_{AR}$
Mean	0.100	0.795	0.403	0.295	0.300	0.100	0.797	0.404	0.296	0.298
SE	0.021	0.085	0.052	0.065	0.034	0.019	0.076	0.052	0.063	0.033
MSE	0.0003	0.006	0.002	0.004	0.001	0.0003	0.005	0.002	0.004	0.001

and AR techniques per 100 iterations makes the former the better estimating option.

Furthermore, Table 10.8 depicts that the model RMSE's of the estimation techniques presented in this book chapter outperform similar benchmarks due to the traditional maximum likelihood estimation (mle) technique given in Ling and Li (1997) for a $ARFIMA(0, d_*, 0) - GARCH(r, s)$ series.

Remark 10.2. Estimation results of a nonstationary ARFIMA(0,d_*,0)-GARCH(1,1) series is continued in the next table.

Due to the positive features of the GARMA(0, d, 0)-GARCH(1,1) model introduced in this chapter, real applications governed by it are provided in the next section.

TABLE 10.2

QMLE estimates for stationary case with d = 0.3, u = 0.8, $\alpha_0 = 0.4$, $\alpha_1 = 0.3$, $\beta = 0.3$, and $v = 1000$

Estimator	$\hat{d}_{MA}$	$\hat{u}_{MA}$	$\hat{\alpha}_{0MA}$	$\hat{\alpha}_{1MA}$	$\hat{\beta}_{MA}$	$\hat{d}_{AR}$	$\hat{u}_{AR}$	$\hat{\alpha}_{0AR}$	$\hat{\alpha}_{1AR}$	$\hat{\beta}_{AR}$
					$n = 100$					
Mean	0.306	0.788	0.430	0.264	0.296	0.305	0.792	0.424	0.273	0.293
SE	0.082	0.130	0.149	0.180	0.098	0.081	0.126	0.148	0.184	0.096
MSE	0.006	0.017	0.023	0.033	0.009	0.006	0.016	0.022	0.034	0.009
					$n = 200$					
Mean	0.301	0.810	0.422	0.276	0.297	0.299	0.800	0.424	0.277	0.292
SE	0.057	0.091	0.136	0.165	0.090	0.060	0.084	0.135	0.161	0.086
MSE	0.003	0.008	0.019	0.027	0.008	0.003	0.007	0.018	0.026	0.007
					$n = 500$					
Mean	0.297	0.814	0.415	0.281	0.299	0.300	0.800	0.415	0.281	0.299
SE	0.037	0.060	0.103	0.123	0.063	0.037	0.051	0.103	0.123	0.063
MSE	0.001	0.003	0.010	0.015	0.004	0.001	0.002	0.010	0.015	0.004
					$n = 1000$					
Mean	0.296	0.814	0.403	0.296	0.300	0.299	0.804	0.406	0.291	0.299
SE	0.026	0.040	0.068	0.085	0.045	0.027	0.033	0.069	0.087	0.045
MSE	0.0007	0.001	0.004	0.007	0.002	0.0007	0.001	0.004	0.007	0.002
					$n = 2000$					
Mean	0.299	0.804	0.403	0.296	0.300	0.299	0.804	0.403	0.296	0.299
SE	0.018	0.022	0.050	0.063	0.034	0.020	0.028	0.052	0.065	0.034
MSE	0.0004	0.0008	0.002	0.004	0.001	0.0003	0.0005	0.002	0.004	0.001

10.7 Forecasting Time Series Using ARMA-GARCH in R – Example

The rugarch package in R is used to estimate and forecast a time series. Estimation of an ARMA model is done initially in R as follows: y <- read-RDS("y.rds")

y.test <- readRDS("y-test.rds")

m1.mean.model <- auto.arima(y, allowmean=F)

ar.comp <- arimaorder(m1.mean.model)[1]

ma.comp <- arimaorder(m1.mean.model)[3]

After running the above R code, it will be evident that usually innovations (residuals or errors) show typical characteristics of a GARCH process.

Subsequently, an ARMA-GARCH(1,1) process is fitted to the data and a one-step ahead forecast on the test data is done by running the following code in R.

TABLE 10.3

QMLE estimates for stationary case with d = 0.45, u = 0.8, $\alpha_0 = 0.4$, $\alpha_1 = 0.3$, $\beta = 0.3$, and $v = 1000$

n = 100										
Estimator	$\hat{d}_{MA}$	$\hat{u}_{MA}$	$\hat{\alpha}_{0MA}$	$\hat{\alpha}_{1MA}$	$\hat{\beta}_{MA}$	$\hat{d}_{AR}$	$\hat{u}_{AR}$	$\hat{\alpha}_{0AR}$	$\hat{\alpha}_{1AR}$	$\hat{\beta}_{AR}$
Mean	0.441	0.820	0.422	0.277	0.295	0.436	0.809	0.427	0.268	0.300
SE	0.058	0.083	0.152	0.183	0.098	0.061	0.087	0.149	0.184	0.100
MSE	0.003	0.007	0.023	0.034	0.009	0.004	0.007	0.023	0.034	0.01
n = 200										
Estimator	$\hat{d}_{MA}$	$\hat{u}_{MA}$	$\hat{\alpha}_{0MA}$	$\hat{\alpha}_{1MA}$	$\hat{\beta}_{MA}$	$\hat{d}_{AR}$	$\hat{u}_{AR}$	$\hat{\alpha}_{0AR}$	$\hat{\alpha}_{1AR}$	$\hat{\beta}_{AR}$
Mean	0.442	0.827	0.422	0.276	0.297	0.444	0.811	0.422	0.276	0.297
SE	0.047	0.057	0.136	0.165	0.090	0.046	0.056	0.136	0.165	0.090
MSE	0.002	0.004	0.019	0.027	0.008	0.002	0.003	0.019	0.027	0.008
n = 500										
Estimator	$\hat{d}_{MA}$	$\hat{u}_{MA}$	$\hat{\alpha}_{0MA}$	$\hat{\alpha}_{1MA}$	$\hat{\beta}_{MA}$	$\hat{d}_{AR}$	$\hat{u}_{AR}$	$\hat{\alpha}_{0AR}$	$\hat{\alpha}_{1AR}$	$\hat{\beta}_{AR}$
Mean	0.453	0.809	0.408	0.290	0.300	0.449	0.808	0.421	0.276	0.301
SE	0.036	0.053	0.089	0.110	0.057	0.033	0.029	0.099	0.123	0.063
MSE	0.001	0.002	0.008	0.012	0.003	0.001	0.0009	0.010	0.015	0.004
n = 1000										
Estimator	$\hat{d}_{MA}$	$\hat{u}_{MA}$	$\hat{\alpha}_{0MA}$	$\hat{\alpha}_{1MA}$	$\hat{\beta}_{MA}$	$\hat{d}_{AR}$	$\hat{u}_{AR}$	$\hat{\alpha}_{0AR}$	$\hat{\alpha}_{1AR}$	$\hat{\beta}_{AR}$
Mean	0.443	0.831	0.406	0.292	0.301	0.449	0.809	0.405	0.293	0.300
SE	0.025	0.022	0.067	0.083	0.045	0.024	0.020	0.068	0.085	0.045
MSE	0.0006	0.001	0.004	0.007	0.002	0.0006	0.0004	0.004	0.007	0.002
n = 2000										
Estimator	$\hat{d}_{MA}$	$\hat{u}_{MA}$	$\hat{\alpha}_{0MA}$	$\hat{\alpha}_{1MA}$	$\hat{\beta}_{MA}$	$\hat{d}_{AR}$	$\hat{u}_{AR}$	$\hat{\alpha}_{0AR}$	$\hat{\alpha}_{1AR}$	$\hat{\beta}_{AR}$
Mean	0.454	0.797	0.403	0.295	0.300	0.450	0.808	0.401	0.299	0.298
SE	0.020	0.018	0.052	0.065	0.034	0.018	0.013	0.052	0.064	0.035
MSE	0.0004	0.00003	0.002	0.004	0.001	0.0003	0.0002	0.002	0.004	0.001

```
library(rugarch)
model.garch = ugarchspec(mean.model=list(armaOrder=c(ar.comp,ma.comp)),
variance.model=list(garchOrder=c(1,1)),
distribution.model = "std")
model.garch.fit = ugarchfit(data=c(y,y.test), spec=model.garch, out.sample
= length(y.test), solver = 'hybrid' ) modelfor=ugarchforecast(model.garch.fit,
data = NULL, n.ahead = 1, n.roll
= length(y.test), out.sample = length(y.test))
results1 <- modelfor@forecastseriesFor[1,] + modelfor@forecastsigmaFor[1,]
results2 <- modelfor@forecastseriesFor[1,] - modelfor@forecastsigmaFor[1,]
ylim <- c(min(y.test), max(y.test))
plot.ts(y.test , col="blue", ylim=ylim)
par(new=TRUE)
modelfor@forecastseriesFor[1,]
par(new=TRUE)
```

TABLE 10.4

Approximation estimation validation results for d = 0.3, u = 0.8, $\alpha_0 = 0.4$, $\alpha_1 = 0.3$, $\beta = 0.3$, and $v = 1000$

			$n = 100$		
m(MA)	$TTP-$ $MSE(MA)$	$Likelihood$ (MA)	m(AR)	$TTP-$ $MSE(AR)$	$Likelihood$ (AR)
6	0.091	-140.917	9	0.090	-122.205
7	0.090	-140.851	10	0.089	-122.184
8	0.093	-140.858	11	0.091	-122.364
			$n = 200$		
m(MA)	$TTP-$ $MSE(MA)$	$Likelihood$ (MA)	m(AR)	$TTP-$ $MSE(AR)$	$Likelihood$ (AR)
6	0.067	-300.327	10	0.067	-248.425
7	0.066	-299.843	11	0.063	-248.208
8	0.069	-300.805	12	0.065	-248.543
			$n = 500$		
m(MA)	$TTP-$ $MSE(MA)$	$Likelihood$ (MA)	m(AR)	$TTP-$ $MSE(AR)$	$Likelihood$ (AR)
6	0.037	-762.011	9	0.035	-759.909
7	0.035	-758.499	10	0.034	-757.427
8	0.037	-858.307	11	0.034	-758.236
			$n = 1000$		
m(MA)	$TTP-$ $MSE(MA)$	$Likelihood$ (MA)	m(AR)	$TTP-$ $MSE(AR)$	$Likelihood$ (AR)
6	0.017	-1484.006	10	0.016	-1406.632
7	0.016	-1478.803	11	0.016	-1406.320
8	0.017	-1479.803	12	0.016	-1406.493
			$n = 2000$		
m(MA)	$TTP-$ $MSE(MA)$	$Likelihood$ (MA)	m(AR)	$TTP-$ $MSE(AR)$	$Likelihood$ (AR)
5	0.010	-2900.998	10	0.009	-2890.806
6	0.009	-2894.747	11	0.008	-2885.443
7	0.009	-2895.531	12	0.009	-2886.006

```
plot.ts(results1, col="red", ylim=ylim)
par(new=TRUE)
plot.ts(results2, col="red", ylim=ylim)
```

Running the above R code would generate the following plots (Figures 10.2 and 10.3).

10.8   Results of Applications

Application 1– S & P 500 Daily Stock index: The log of the daily S and P 500 index with 8435 observations from 2nd January 1980 to 10th June

TABLE 10.5

Optimal values of m for varying d and u using both approximations with 1000 replications

n	$d=0.1$ (MA)	$d=0.3$ (MA)	$d=0.45$ (MA)	$d=0.1$ (AR)	$d=0.3$ (AR)	$d=0.45$ (AR)
100	6	7	7	9	10	10
200	7	7	7	9	11	11
500	8	7	6	9	10	11
1000	8	7	7	11	11	11
2000	9	6	6	12	11	11

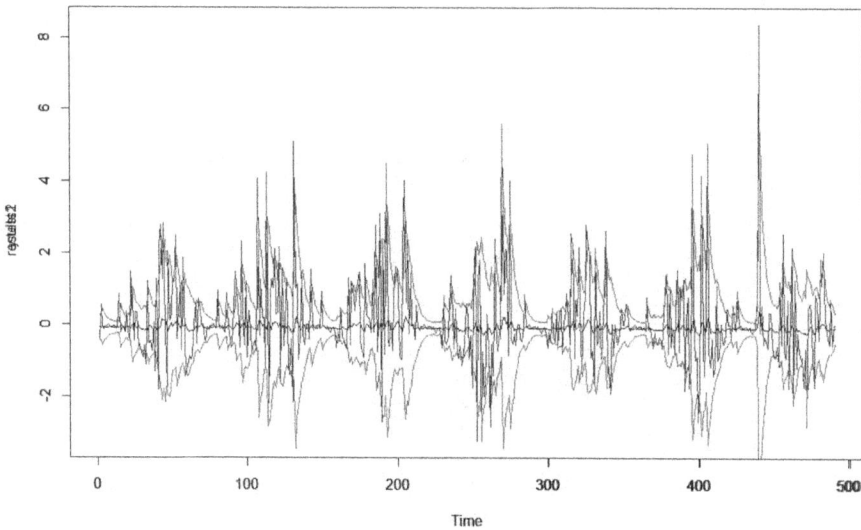

FIGURE 10.2

Plot 1 generated by code.

2013 [Source: https://au.finance.yahoo.com] was used. A large time series of the Daily S and P 500 index was chosen as a real application of the introduced process anticipating the persistence of conditional heteroskedasticity. The acf, pacf and sdf of the series did suggest that it has generalized persistence, since the acf and the pacf, depict hyperbolic decays and the sdf consists of unbounded peaks away from the origin. In lieu of it a GARMA$(0, d, 0)$-GARCH$(1,1)$ model was suggested and fitted using the better option of the MA approximation.

The application yielded the following results at an optimal truncation lag order $m = 6$ (within optimal lag order interval): $\hat{d}_{MA} = 0.1042$, $\hat{u}_{MA} = 0.9151$, $\hat{\alpha}_{0MA} = 2.6422 \times 10^{-7}$, $\hat{\alpha}_{1MA} = 0.078939$, $\hat{\beta}_{MA} = 0.91106$.

TABLE 10.6

QMLE estimates for nonstationary ARFIMA$(0,d_*,0)$-GARCH$(1,1)$ series

$n = 200$, $d_* = 0.7$, $\alpha_0 = 0.3$, $\alpha_1 = 0.3$, $\beta = 0.3$, $v = 500$							
Estimator	$\hat{d}_{*MA}$	$\hat{\alpha}_{0MA}$	$\hat{\alpha}_{1MA}$	$\hat{\beta}_{MA}$	$\hat{d}_{*AR}$	$\hat{\alpha}_{0AR}$	$\hat{\alpha}_{1AR}$	$\hat{\beta}_{AR}$
Mean	0.7009	0.3216	0.2667	0.2980	0.7067	0.3120	0.2805	0.3005
SE	0.0631	0.1130	0.1823	0.0980	0.0041	0.1200	0.1819	0.0988
MSE	0.0040	0.0132	0.0343	0.0096	0.000060893	0.0148	0.0335	0.0098
$n = 400$, $d_* = 0.7$, $\alpha_0 = 0.3$, $\alpha_1 = 0.3$, $\beta = 0.3$, $v = 500$							
Estimator	$\hat{d}_{*MA}$	$\hat{\alpha}_{0MA}$	$\hat{\alpha}_{1MA}$	$\hat{\beta}_{MA}$	$\hat{d}_{*AR}$	$\hat{\alpha}_{0AR}$	$\hat{\alpha}_{1AR}$	$\hat{\beta}_{AR}$
Mean	0.7070	0.3120	0.2841	0.2972	0.7068	0.3120	0.2805	0.3005
SE	0.0012	0.0942	0.1501	0.0774	0.0025	0.0886	0.1458	0.0763
MSE	0.000050056	0.0090	0.0227	0.0060	0.000052426	0.0080	0.0216	0.0058
$n = 200$, $d_* = 1.0$, $\alpha_0 = 0.2$, $\alpha_1 = 0.2$, $\beta = 0.2$, $v = 500$							
Estimator	$\hat{d}_{*MA}$	$\hat{\alpha}_{0MA}$	$\hat{\alpha}_{1MA}$	$\hat{\beta}_{MA}$	$\hat{d}_{*AR}$	$\hat{\alpha}_{0AR}$	$\hat{\alpha}_{1AR}$	$\hat{\beta}_{AR}$
Mean	0.9947	0.2009	0.2022	0.1905	0.9951	0.2009	0.2022	0.1905
SE	0.0250	0.0696	0.2164	0.0817	0.0241	0.0696	0.2164	0.0817
MSE	0.0006498	0.0048	0.0468	0.0067	0.00060478	0.0048	0.0468	0.0067
$n = 400$, $d_* = 1.0$, $\alpha_0 = 0.2$, $\alpha_1 = 0.2$, $\beta = 0.2$, $v = 500$							
Estimator	$\hat{d}_{*MA}$	$\hat{\alpha}_{0MA}$	$\hat{\alpha}_{1MA}$	$\hat{\beta}_{MA}$	$\hat{d}_{*AR}$	$\hat{\alpha}_{0AR}$	$\hat{\alpha}_{1AR}$	$\hat{\beta}_{AR}$
Mean	0.9986	0.1981	0.2108	0.1940	0.9984	0.1981	0.2108	0.1940
SE	0.0124	0.0615	0.1949	0.0687	0.0131	0.0615	0.1949	0.0687
MSE	0.00015565	0.0038	0.0380	0.0048	0.00017448	0.0038	0.0380	0.0048
$n = 200$, $d_* = 1.2$, $\alpha_0 = 0.25$, $\alpha_1 = 0.25$, $\beta = 0.25$, $v = 500$							
Estimator	$\hat{d}_{*MA}$	$\hat{\alpha}_{0MA}$	$\hat{\alpha}_{1MA}$	$\hat{\beta}_{MA}$	$\hat{d}_{*AR}$	$\hat{\alpha}_{0AR}$	$\hat{\alpha}_{1AR}$	$\hat{\beta}_{AR}$
Mean	1.2045	0.2617	0.2294	0.2419	1.2030	0.2617	0.2294	0.2419
SE	0.0644	0.0912	0.1956	0.0914	0.0659	0.0912	0.1956	0.0914
MSE	0.0042	0.0084	0.0386	0.0084	0.0043	0.0084	0.0386	0.0084
$n = 400$, $d_* = 1.2$, $\alpha_0 = 0.25$, $\alpha_1 = 0.25$, $\beta = 0.25$, $v = 500$							
Estimator	$\hat{d}_{*MA}$	$\hat{\alpha}_{0MA}$	$\hat{\alpha}_{1MA}$	$\hat{\beta}_{MA}$	$\hat{d}_{*AR}$	$\hat{\alpha}_{0AR}$	$\hat{\alpha}_{1AR}$	$\hat{\beta}_{AR}$
Mean	1.2184	0.2547	0.2444	0.2437	1.2175	0.2547	0.2444	0.2437
SE	0.0395	0.0791	0.1752	0.0749	0.0426	0.0791	0.1752	0.0749
MSE	0.0019	0.0063	0.0307	0.0056	0.0021	0.0063	0.0307	0.0056

Therefore, the fitted $GARMA(0,d,0) - GARCH(r,s)$ model is:

$$(1 - 2 \times 0.9151B + B^2)^{0.1042} X_t = \varepsilon_t,$$

where the standard errors of $\hat{d}_{MA}$, $\hat{u}_{MA}$, $\hat{\alpha}_{0MA}$, $\hat{\alpha}_{1MA}$, $\hat{\beta}_{MA}$ are 2.3131×10^{-7}, 0.0763, 2.3398×10^{-8}, 0.0018024, and 0.0029494, respectively.

An in-sample rolling forecast was performed to assess the final 200 observations of the data set resulting in a one-step ahead forecast MSE of 2.4723×10^{-5}. Original and forecast values of the final 200 observations did depict that they were reasonably close, proving that the utilized model was feasible in terms of forecast accuracy.

Application 2– Chicago Board Options Exchange (CBOI) market expectation daily measurement volatility index: As a second application, the log of

TABLE 10.7

QMLE estimates for nonstationary ARFIMA(0,d_*,0)-GARCH(1,1) series

	$n = 200$, $d_* = 1.4$, $\alpha_0 = 0.3$, $\alpha_1 = 0.3$, $\beta = 0.3$, $v = 500$							
Estimator	$\hat{d}_{*MA}$	$\hat{\alpha}_{0MA}$	$\hat{\alpha}_{1MA}$	$\hat{\beta}_{MA}$	$\hat{d}_{*AR}$	$\hat{\alpha}_{0AR}$	$\hat{\alpha}_{1AR}$	$\hat{\beta}_{AR}$
Mean	1.3757	0.3209	0.2769	0.2899	1.3654	0.3117	0.2808	0.2997
SE	0.0934	0.1200	0.1819	0.0988	0.0987	0.1063	0.1807	0.0992
MSE	0.0093	0.0148	0.0335	0.0098	0.0109	0.0114	0.0330	0.0098
	$n = 400$, $d_* = 1.4$, $\alpha_0 = 0.3$, $\alpha_1 = 0.3$, $\beta = 0.3$, $v = 500$							
Estimator	$\hat{d}_{*MA}$	$\hat{\alpha}_{0MA}$	$\hat{\alpha}_{1MA}$	$\hat{\beta}_{MA}$	$\hat{d}_{*AR}$	$\hat{\alpha}_{0AR}$	$\hat{\alpha}_{1AR}$	$\hat{\beta}_{AR}$
Mean	1.4032	0.3120	0.2841	0.2972	1.4038	0.3202	0.2693	0.2992
SE	0.0542	0.0942	0.1501	0.0774	0.0493	0.0973	0.1548	0.0788
MSE	0.0029	0.0090	0.0227	0.0060	0.0024	0.0099	0.0249	0.0062
	$n = 200$, $d_* = 2.2$, $\alpha_0 = 0.2$, $\alpha_1 = 0.2$, $\beta = 0.2$, $v = 500$							
Estimator	$\hat{d}_{*MA}$	$\hat{\alpha}_{0MA}$	$\hat{\alpha}_{1MA}$	$\hat{\beta}_{MA}$	$\hat{d}_{*AR}$	$\hat{\alpha}_{0AR}$	$\hat{\alpha}_{1AR}$	$\hat{\beta}_{AR}$
Mean	2.2288	0.2001	0.2034	0.1943	2.2202	0.2009	0.2022	0.1905
SE	0.0458	0.0680	0.2162	0.0886	0.0541	0.0696	0.2164	0.0817
MSE	0.0029	0.0046	0.0467	0.0079	0.0033	0.0048	0.0468	0.0067
	$n = 400$, $d_* = 2.2$, $\alpha_0 = 0.2$, $\alpha_1 = 0.2$, $\beta = 0.2$, $v = 500$							
Estimator	$\hat{d}_{*MA}$	$\hat{\alpha}_{0MA}$	$\hat{\alpha}_{1MA}$	$\hat{\beta}_{MA}$	$\hat{d}_{*AR}$	$\hat{\alpha}_{0AR}$	$\hat{\alpha}_{1AR}$	$\hat{\beta}_{AR}$
Mean	2.2342	0.1981	0.2108	0.1940	2.2326	0.1981	0.2108	0.1940
SE	0.0152	0.0615	0.1949	0.0687	0.0263	0.0615	0.1949	0.0687
MSE	0.0014	0.0038	0.0380	0.0048	0.0018	0.0038	0.0380	0.0048

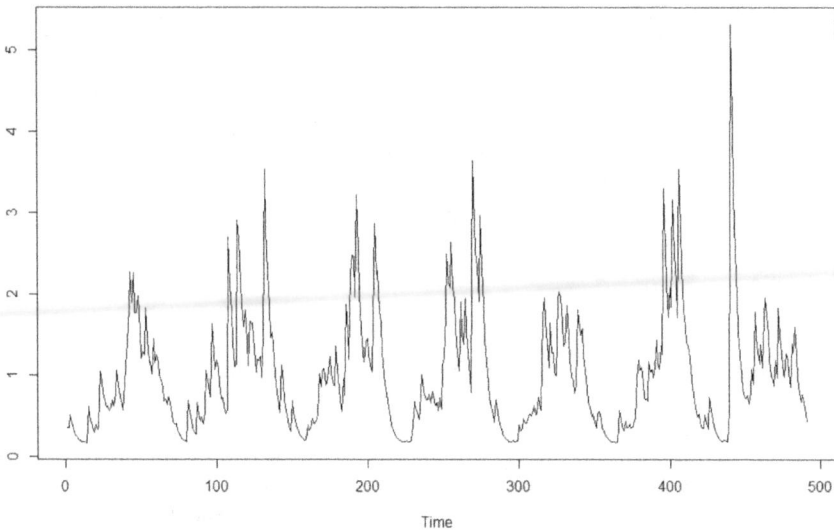

FIGURE 10.3

Plot 2 generated by code.

TABLE 10.8

TTP-RMSE estimate comparison table for nonstationary ARFIMA(0,d_*,0)-GARCH(1,1) series

$n = 200$, $d_* = 0.7$, $\alpha_0 = 0.3$, $\alpha_1 = 0.3$, $\beta = 0.3$, $v = 500$		
Li and Ling - MLE	QMLE(MA Approximation)	QMLE(AR Approximation)
0.498	0.4613	0.4114
$n = 400$, $d_* = 0.7$, $\alpha_0 = 0.3$, $\alpha_1 = 0.3$, $\beta = 0.3$, $v = 500$		
Li and Ling - MLE	QMLE(MA Approximation)	QMLE(AR Approximation)
0.400	0.3300	0.3198
$n = 200$, $d_* = 1.0$, $\alpha_0 = 0.2$, $\alpha_1 = 0.2$, $\beta = 0.2$, $v = 500$		
Li and Ling - MLE	QMLE(MA Approximation)	QMLE(AR Approximation)
0.469	0.3929	0.3920
$n = 400$, $d_* = 1.0$, $\alpha_0 = 0.2$, $\alpha_1 = 0.2$, $\beta = 0.2$, $v = 500$		
Li and Ling - MLE	QMLE(MA Approximation)	QMLE(AR Approximation)
0.414	0.3383	0.3390
$n = 200$, $d_* = 1.2$, $\alpha_0 = 0.25$, $\alpha_1 = 0.25$, $\beta = 0.25$, $v = 500$		
Li and Ling - MLE	QMLE(MA Approximation)	QMLE(AR Approximation)
0.490	0.4445	0.4453
$n = 400$, $d_* = 1.2$, $\alpha_0 = 0.25$, $\alpha_1 = 0.25$, $\beta = 0.25$, $v = 500$		
Li and Ling - MLE	QMLE(MA Approximation)	QMLE(AR Approximation)
0.433	0.3730	0.3752
$n = 200$, $d_* = 1.4$, $\alpha_0 = 0.3$, $\alpha_1 = 0.3$, $\beta = 0.3$, $v = 500$		
Li and Ling - MLE	QMLE(MA Approximation)	QMLE(AR Approximation)
0.551	0.5001	0.4918
$n = 400$, $d_* = 1.4$, $\alpha_0 = 0.3$, $\alpha_1 = 0.3$, $\beta = 0.3$, $v = 500$		
Li and Ling - MLE	QMLE(MA Approximation)	QMLE(AR Approximation)
0.417	0.3768	0.3850
$n = 200$, $d_* = 2.2$, $\alpha_0 = 0.2$, $\alpha_1 = 0.2$, $\beta = 0.2$, $v = 500$		
Li and Ling - MLE	QMLE(MA Approximation)	QMLE(AR Approximation)
0.501	0.4266	0.4249
$n = 400$, $d_* = 2.2$, $\alpha_0 = 0.2$, $\alpha_1 = 0.2$, $\beta = 0.2$, $v = 500$		
Li and Ling - MLE	QMLE(MA Approximation)	QMLE(AR Approximation)
0.422	0.3632	0.3682

the daily CBOI index with 2744 observations from 2nd January 2004 to 10th November 2014 [Source: http://www.cboe.com] was used in fitting a GARMA(0, d, 0)-GARCH(1,1) model using the better option of the MA approximation, since the acf, pacf, and the sdf did suggest generalized persistence quite similar to the corresponding functions of the S & P daily stock index of application 1.

The application yielded the following results at an optimal truncation lag order $m = 6$ (within optimal lag order interval): $\hat{d}_{MA} = 0.1196$, $\hat{u}_{MA} = 0.7932$, $\hat{\alpha}_{0MA} = 7.6094 \times 10^{-5}$, $\hat{\alpha}_{1MA} = 0.11957$, $\hat{\beta}_{MA} = 0.79317$.

Therefore, the fitted $GARMA(0, d, 0) - GARCH(r, s)$ model is
$$(1 - 2 \times 0.7932B + B^2)^{0.1196} X_t = \varepsilon_t,$$
where the standard errors of $\hat{d}_{MA}$, $\hat{u}_{MA}$, $\hat{\alpha}_{0MA}$, $\hat{\alpha}_{1MA}$, and $\hat{\beta}_{MA}$ are 7.6089×10^{-5}, 0.0719, 8.4106×10^{-6}, 0.011141, and 0.017216, respectively.

In a theoretical sense for both applications, the values of $\hat{d}_{MA}$, $\hat{u}_{MA}$, $\hat{\alpha}_{0MA}$, $\hat{\alpha}_{1MA}$, and $\hat{\beta}_{MA}$ fall within the stipulated bounds of a $GARMA(0,d,0) - GARCH(r,s)$ series as per the model definitions of (10.2), (10.3) and (10.5). Therefore, the chosen daily S and P 500 and CBOI index series theoretically illustrate the properties of heteroskedasticity and persistence. It results in $GARMA(0,d,0) - GARCH(1,1)$ models with a high degree of volatility due to conditional heteroskedasticity.

Based on the positive attributes of the introduced model and contributions of this chapter illustrated in Sections 1-4 concluding remarks are provided in the next section.

10.9   Discussion

A truncated state space model entailing the KF of a simple Gegenbauer process is introduced to arrive at two types of parameter estimates. These two types are constructed using AR and MA approximations of the series to introduce optimal lag order intervals. A comparative assessment of the two approximations based on Monte Carlo evidence was provided next as an interesting proposition. It also did prove that in the nonstationary form of the model the two introduced approximate QMLE techniques are superior to a traditional MLE mechanism in the literature. Finally, the better approximation method is applied to two real applications within the established optimal lag order interval. It proves the persistence of volatility as per the introduced model within a long-duration interval. Therefore, the introduced process could significantly impact the global economy if a large volume of data indices behave in line with it.

10.10   Chapter 10 Questions

1. Differentiate white noise with GARCH errors?
2. Develop a state space configuration for an ARIMA(0,d,0)-GARCH(1,1) model using the state space model presented in this chapter for a GARMA(0,d,0)-GARCH(1,1) model?
3. Develop a state space configuration for an ARMA(1,1)-GARCH(1,1) model using the state space model presented in this chapter for a GARMA(0,d,0)-GARCH(1,1) model?
4. Run the following R code and assess the output?

    set.seed(1)

    require(TSA)

    t=numeric(400)

    x=garch.sim(alpha=c(0.2,0.2),beta=c(0.2),n=1000)

    y=x[201:1000]

```
p=numeric(400)
coeff=numeric(800)
g=numeric(400)
for(j in 3:800)
l=0.45
v=0.8
coeff[1]=1
coeff[2]=0.72
coeff[j]=(((2*v)*(j+l-1)*coeff[j-1])-((j+(2*l)-2)*coeff[j-2]))/j
cone=coeff[1:400]
ctwo=coeff[401:800]
k=1:400
for(i in 401:800)
g[i]=sum(cone[k]*y[i-k])
t=g[401:800]
for(n in 1:400)
p[n]=sum(cone[k]*ctwo[n])
Realization=ts.plot(t)olatility
par(mfrow=c(2,2))
ts.plot(t)
acf(t)
acf(p)
acf(t,type="partial")
dev.off()
```

5. Compare the figures generated by running the code in 4 above with the figures provided in this book chapter. Are they similar?

6. Run the same code given in 4 with the fourth line command syntax changed to: x=garch.sim(alpha=c(0.3,0.3),beta=c(0.3),n=1000) and obtain the solution?

7. What's the difference between the terms **volatility** and **heteroskedasticity** ?

8. Develop an R code (similar to the one provided in question 4) for a stationary ARFIMA(1,d,1)-GARCH(1,1) model?

9. Develop an R code (similar to the one provided in question 4) for a nonstationary ARFIMA(1,d,1)-GARCH(1,1) model?

10. From the results of questions 8 and 9 perform a comparative assessment of stationary and nonstationary ARFIMA(1,d,1)-GARCH(1,1) models?

11

Enhancing Time Series Analysis with Machine Learning, High-Frequency Data, and Applications in Medicine and Biology

Synopsis: This chapter on time series analysis utilizing machine learning (ML) would add supplementary novelty and flavor to preceding chapters. An introduction to the topic involving the usage of the tsfGRNN (time series forecasting using general regression neural network) package in this chapter would add an extra dimension to time series analysis in terms of knowledge; especially in terms of using high-frequency data.

11.1 Introduction

Time series forecasting (predicting or projecting) is a common task in many fields such as business, medicine, public health, agriculture, or environment. Effective prediction estimates could optimize time and money, incorporating planning, scheduling, and other activities.

There are scenarios where a great number of time series are needed forecast quickly, such as the selling of different retail products. In such a context speedy robust forecasting tools, which can be applied with little or no human interference, are highly valuable. These requirements preclude some common methodologies; for example, ARIMA methodology is usually applied under expert supervision, while other projection tools can be used with minimal human intervention but have high computational needs.

11.2 Preliminaries

ML is the use and development of computer systems that are able to learn and adapt without following explicit instructions, by using algorithms and statistical models to analyze and draw inferences from patterns in data. It is a subject that has three main subdivisions as follows:

(1) Supervised Learning – use and development of computer systems that are able to learn and adapt by following explicit instructions, by using

algorithms and statistical models to analyze and draw inferences from patterns in data.

(2) Unsupervised Learning – use and development of computer systems that are able to learn and adapt without following explicit instructions, by using algorithms and statistical models to analyze and draw inferences from patterns in data.

(3) Reinforced Learning – use and development of computer systems that are able to learn and adapt by following instructions from the past, by using algorithms and statistical models to analyze and draw inferences from patterns in data.

High-frequency data refer to time-series data collected at a very fine scale, often at intervals of seconds or even milliseconds. It's commonly used in fields like finance, where understanding market behaviors and microstructures requires observing transactions and quotes at high-frequency.

Note: The "tsfGRNN" package introduced in the next subsection of this chapter is an unsupervised learning technique in ML that could be applied to high-frequency data in time series analysis.

11.3   tsfGRNN Package

Generalized regression neural networks (GRNNs) are a variant of radial basis function networks. They do exhibit interesting properties to develop a fast forecasting tool as follows:

(1) They possess single-pass learning,

(2) They only need to either set or fit one parameter, and

(3) They do produce deterministic results, so that it is not required to train several neural networks to achieve more accurate and trustworthy outcomes.

Therefore, the combining of time series forecasting methodologies with GRNN networks in ML did result in the advent of the "time series forecasting generalized regression neural networks (tsfGRNN) package. An application example of it is illustrated in the next subsection.

11.4   An Illustrative Example

The tsfgrnn package is a R package for time series forecasting using GRNN, which implements different modeling and transforming approaches. To install the package from CRAN the install.packages command can be used at the console by typing install.packages("tsfgrnn") in front of the command prompt.

After installing the tsfGRNN package, you may run the following R code as an exercise as well as a practicing example.

```
> library(tsfGRNN)
> pred < −grnnforecasting(UKgas, h = 4)
> pred$prediction
> plot(pred)
> library(ggplot2)
> autoplot(pred)
```

In the above code, the UK gas data prior to the year 1987 are used and h denotes the number of seasonal steps ahead (quarters per year) in the forecasting horizon. By running the code the resulting output would be predictions on gas output in the United Kingdom for quarters 1-4 in 1987. Refer Martínez et al. (2022) for further clarifications of the material presented in this chapter.

11.5 Discussion

In this chapter, an unsupervised learning technique in ML that is applicable to high-frequency data is introduced and illustrated through a coding example using the R package.

In the recent past time series, methodology has been utilized to assess natural occurrences in medicine and biology. One such branch in medicine happens to be epidemiology that encompasses epidemics and pandemics. Since epidemics and pandemics spread in human society with respect to time, the independent variable of such a process becomes time. In such a context, time series methods could be used to develop both in-sample and out-of-sample forecast projections with respect to disease incidence and prevalence.

In another recent development, a study was done comprising of a daily time-series analysis using a Poisson generalized additive model of heatwaves coded as a binary variable (1 for heatwave day, 0 for nonheatwave day). The model was adjusted for confounders including day of week, humidity and seasonal trends in mortality (refer Chaston et al. (2022)) for further details.

In yet another development, quantity of rainfall and its related events have become more and more uncertain due to climatic variability. The complexity of the rainfall pattern increases due to the changes of the atmospheric behavior from time to time. Relatively, few measures have been taken to perform the modeling of rainfall in the context of long memory. A recent study in applied statistics provides an assessment of such a phenomenon by fitting an appropriate time series model. A longrange dependency model is proposed to fit weekly rainfall data to explore characteristics of persistence through an unbounded spectral density. Careful examination of the data exhibits periodic fluctuations as an additional feature. Since the rainfall series exhibits periodic variations and persistence, a seasonal autoregressive fractionally integrated moving average (SARFIMA) model is fitted. Parameters of it are estimated using maximum likelihood estimation (MLE) method. A Monte Carlo simulation was carried out with different seasonal and nonseasonal fractionally differing parameters to measure the suitability of the method for parameter estimation. Best-fitted model was chosen based on the minimum of the mean absolute error and the forecasting performance and are compared with the result of a seasonal autoregressive integrated moving average (SARIMA) using an independent sample as a new applicable study in time series analysis.

11.6 Chapter 11 Questions

1. What is ML?
2. What's high-frequency data?
3. What are the three main subdivisions of ML?
4. What type of a package is "tsfGRNN"?
5. Produce the output by running the R code presented in this chapter for h = 12 (months per year)?
6. Produce the output by running the R code presented in this chapter for h = 365 (days per year)?
7. Produce the output by running the R code presented in this chapter for h = 12 (months per year) for a different embedded data set in the CRAN repository?
8. Produce the output by running the R code presented in this chapter for h = 365 (days per year) for a different embedded data set in the CRAN repository?
9. Produce the output by running the R code presented in this chapter for h = 4 (quarters per year) for a different embedded data set in the CRAN repository?
10. Try to provide interpretations for the plots generated in exercises 5–9?
11. What type of features could be found in a basic SARIMA model?
12. What type of features could be found in a basic SARFIMA model?
13. Why does epidemic or pandemic incidence frequency data categorized as time series data?
14. What characteristics could be used to distinguish between SARIMA and SARFIMA time series models?
15. In this chapter what are the reasons for fitting a long-range dependency model to fit weekly rainfall data ?

Bibliography

Abrahams, M. and Dempster, A. (1979) Research on seasonal analysis. progress report on the asa census project on seasonal adjustment. Technical report. Department of Statistics, Harvard University, Boston, MA.

Andel, J. (1986). Long memory time series models. Kybernetika, 22(2), 105-123.

Andersen, T. G., Bollerslev, T., Diebold, F. X. and Ebens, H. (2001a). The distribution of realized stock return volatility. Journal of Financial Economics, 61, 43-76.

Andersen, T. G., Bollerslev, T., Diebold, F. X. and Labys, P. (2001b). The distribution of realized exchange rate volatility. Journal of American Statistical Association, 96, 42-55.

Anderson, B. D. O. and Moore, J. B. (1979). Optimal Filtering. Prentice-Hall, New York. Anh, V. V.

Aoki, M. (1990). State Space Modeling of Time Series. Springer, Berlin.

Arteche, J. and Robinson, P. M. (2000). Semiparametric inference in seasonal and cyclical long memory processes. Journal of Time Series Analysis, 21, 1-25.

Arteche, J. (2007). The Analysis of Seasonal Long Memory: The Case of Spanish Inflation. 132 BIBLIOGRAPHY Oxford Bulletin of Economics and Statistics, 69, 749-772. 133

Arteche, J. (2012). Standard and Seasonal Long Memory in Volatility: An application to Spanish Inflation. Empirical Economics, 42, 693-712.

Baillie, R. T. (1996). Long Memory Processes and Fractional Integration in Econometrics. Journal of Econometrics, 73, 5-59.

Baillie, R. T., Bollerslev, T. and Mikkelsen, H. O. (1996a). Fractionally Integrated Generalized Autoregressive Conditional Heteroskedasticity. Journal of Econometrics, 74(1), 3-30.

Baillie, R. T.,Chung, C. F. and Tieslau, M. A. (1996). Analysing inflation by the fractionally integrated ARFIMA-GARCH model. Journal of Applied Econometrics, 11, 23-40.

Beaumont, P. and Ramachandran, R. (2001). Robust Estimation of GARMA Model Parameters with an Application to Cointegration among Interest Rates of Industrialized Countries. Computational Economics, 17, 179-201.

Bengio, Y., Courville, A. and Goodfellow, I. (2016). Deep Learning, MIT Press.

Beran, J. (1994). Statistics for Long-Memory Processes. Chapman and Hall,New York.

Beran, J., Feng, Y., Ghosh, S. and Kulik, R. (2013). Long-Memory Processes Probabilistic Properties and Statistical Methods, Springer-Verlag Berlin Heidelberg.

Bickel, P. J. and Doksum, K. A. (2001). Mathematical Statistics: Basic Ideas and Selected Topics, Pearson.

Bijari, M. and Khashei, M. (2011). A novel hybridization of artificial neural networks and ARIMA models for time series forecasting. Applied Soft Computing, 11 (2011), 2664-2675.

Bisaglia, L., Bordignon, S. and Lisi, F. (2003). K-Factor GARMA models for intraday volatility forecasting. Applied Economics Letters, 10(4), 256-264.

Bisognin, C. and Lopes, S. R. C. (2009). Properties of Seasonal Long Memory Processes. Mathematical and Computer Modeling, 49, 1837-1851.

Bollerslev, T. (1986). Generalized Autoregressive Conditional Heteroskedasticity. Journal of Econometrics, 31, 307-327.

Bos, C. S., Koopman, S. J. and Ooms, M. (2014). Long Memory with Stochastic Variance Model: A Recursive Analysis for US Inflation. Computational Statistics and Data AnalysisThe Annals of Computational and Financial Econometrics- Issue 2, 76, 144-157.

Box, G. E. P. and Jenkins, G. M. (1970). Time Series Analysis: Forecasting and Control. San Francisco: Holden-Day.

Brockwell, P. J. and Davis, R. A. Time Series: Theory and Methods. Springer-Verlag: New York, 1991.

Brockwell, P. J. and Davis, R. A. Introduction to Time Series and Forecasting. Springer: New York, 1996.

Chan, N. H. and Palma, W. (1998). State Space Modeling of Long-Memory Processes. The Annals of Statistics, 26(2), 719-740.

Chan, N. H. and Palma, W. (2006). Estimation of Long-Memory Time Series Models: A Survey of Different Likelihood-Based Methods. Advances in Econometrics, 20(2), 89-121.

Chaston, T. B., Broome, R. A., Cooper, N., Duck, G., Geromboux, C., Guo, Y., Ji, F., Perkins-Kirkpatrick, S., Zhang, Y., Dissanayake, G. S. and Morgan, G. G., 2022. Mortality burden of heatwaves in Sydney, Australia is exacerbated by the urban heat island and climate change: can tree cover help mitigate the health impacts?. Atmosphere, 13(5), p.714.

Chen, G., Abraham, B. and Peiris, M. S. (1994). Lag Window Estimation of the degree of differencing in fractionally integrated time series models. Journal of Time Series Analysis, 15(5), 473-487.

Chung,C.-F.(1996). A Generalized Fractionally Integrated Autoregressive Moving-Average Process. Journal of Time Series Analysis, 17(2), 111-140.

Cleveland, W. S. and Devlin, S. J. (1998). Locally Weighted Regression: An Approach to Regression Analysis by Local Fitting. Journal of the American Statistical Association, 83(403), 596-610.

Corsi, F. (2009). A Simple Approximate Long-Memory model of Realized Volatility. Journal of Financial Econometrics, 7(2), 174-196.

Dahlhaus, R. (1989). Efficient Parameter Estimation for Self Similar Processes. The Annals of Statistics, 17(4), 1749-1766.

Dahlhaus, R. (2000). Nonlinear Time Series: Semiparametric and Nonparametric Methods, Springer.

Diebold, F. X., Husted, S. and Rush, M. (1991). Real exchange rates under the gold standard. Journal of Political Economy, 99, 1252-1271.

Diebold, F. X. and Rudebusch, G. D. (1991). On the Power of the Dickey-Fuller Tests against Fractional Alternatives. Economic Letters, 35, 155-160.

Dissanayake, G. S. and Peiris, M. S. (2011). Generalized Fractional Processes with Conditional Heteroskedasticity. Sri Lankan Journal of Applied Statistics, 12, 1-12.

Dissanayake, G. S., Peiris, M. S. and Proietti. T. (2014a). State Space Modeling of Gegenbauer Processes with Long Memory. Computational Statistics and Data Analysis- The Annals of Computational and Financial Econometrics, http://dx.doi.org/10.1016/j.csda.2014.09.014

Dissanayake G. S., Peiris, M. S. and Proietti, T. (2014b). Estimation of Generalized Fractionally Differenced Processes with Conditionally Heteroskedastic Errors. International Work Conference on Time Series, Proceedings ITISE 2014, Ignacio Rojas Ruiz and Gonzalo Ruiz Garcia (ed.), Copicentro Granada S L, ISBN 978-84-15814-97-9, 871–890.

Dolado, J. J., Gonzalo, J. and Mayoral, L. (2002). A Fractional Dickey-Fuller Test for Unit Roots. Econometrica, 70(5), 1963-2006.

Durbin, J. and Koopman, S. J.(2001). Time Series Analysis by State Space Methods. Number 24 in Oxford Statistical Science Series. Oxford: Oxford University Press.

Engle, R. F. (1982) Autoregressive Conditional Heteroskedasticity with Estimates of Variance of U.K. Inflation. Econometrica, 50, 987-1008.

Erdelyi, A., Magnus, W., Oberhettinger, F., and Tricomi, F. G. (1953). Higher Transcendental Functions, Vol II, Bateman Manuscript Project, McGraw and Hill.

Fan, J. and Gijbels, L. (1996). Local Polynomial Modeling and Its Applications, CRC Press.

Fan, J. and Yao, Q. (2003). Nonlinear Time Series: Nonparametric and Parametric Methods, Springer.

Ferrara, L. and Guegan, D. (2001). Forecasting with k-factor Gegenbauer Processes: Theory and Applications. Journal of Forecasting, 20(8), 581-601.

Ferrara, L., Guegan, D. and Lu, Z. (2010). Testing Fractional Order of Long Memory Processes: A Monte Carlo Study. Communications in Statistics- Simulation and Computation, 39(4), 795-806.

Fox, R. and Taqqu, M. S.(1986). Large Sample Properties of Parameter Estimates for Strongly Dependent Stationary Gaussian Time Series. The Annals of Statistics, 14(2), 517-532.

Geweke, J. and Porter-Hudak, S. (1983). The estimation and application of long Memory time series models. Journal of Time Series Analysis, 4(4), 221-238.

Ghosh, D. and Stoev, S. (2011). Nonparametric Estimation of Long-Memory Time Series Models. Journal of the Royal Statistical Society: Series B (Statistical Methodology), 73(2), 295-315.

Giraitis, L. and Leipus, R. (1995) A Generalized Fractionally Differencing approach in Long Memory modeling. Lithuanian Mathematical Journal, 35(1), 65-81.

Giraitis, L., Koul, H. L. and Surgailis, D. (2012). Large Sample Inference for Long Memory Processes. London: Imperial College Press.

Gould, H. W. (1974) Coefficient Identities for Powers of Taylor and Dirichlet Series. The American Mathematical Monthly, 81(1), 3-14.

Granger, C. W. J. and Joyeux, R. (1980). An introduction to long-memory time series models and fractional differencing. Journal of Time Series Analysis, 1, 15-29.

Grassi, S., and Santucci de Magistris, P. (2014). When long memory meets the Kalman filter: A comparative Study. Computational Statistics and Data Analysis, 76, 301-319.

Gray, H. L., Woodward, W. A. and Zhang, N. F. (1989). On generalized fractional processes. Journal of Time Series Analysis, 10, 233-257.

Guegan, D. (2005). How can we define the Concept of Long Memory? An Econometric Survey. Econometric Reviews, 24(2), 113-149.

Hall, P. and Horowitz, J. L. (2007). Methodology and Applications of Nonparametric Econometrics. Journal of Econometrics, 137(1), 1-12.

Haldrup, N. and Nielsen, M. Ø. (2007). Estimation of fractional integration in the presence of data noise. Computational Statistics & Data Analysis, 51(6), pp.3100-3114.

Hannan, E. J. (1979). The Central Limit Theorem for Time Series Regression. Stochastic Processes and Their Applications, 9, 281-289.

Härdle, W. (1992). Kernel regression smoothing of time series. Journal of Time Series Analysis, 13(3), 209–232. https://doi.org/10.1111/j.1467-9892.1992.tb00103.x

Harvey, A. C. (1989). Forecasting, Structural Time Series Models and the Kalman Filter. Cambridge: Cambridge University Press.

Harvey, A. C. and Proietti, T.(2005). Readings in Unobserved Components Models. Oxford: Oxford University Press.

Hassler, U. and Wolters, J. (1994). On the Power of Unit Root Tests against Fractional Alternatives. Economics Letters, 45, 1-5.

Hassler, U. and Wolters, J. (1995). Long memory in inflation rates: International evidence. Journal of Business and Economic Statistics, 13, 37-45.

Hastie, T. J. and Tibshirani, R. J. (1990). Generalized Additive Models, CRC Press.

Ho, K. W. and Houmani, H. (2010). Investigation of GARCH models for the Estimation Power and Normality. Master's Thesis. Lund University.

Hosking, J. R. M. (1981). Fractional Differencing. Biometrika, 68, 165-176. 139

Hsu, N.- J. and Tsai, H. (2009) Semiparametric Estimation for Seasonal Long Memory Time Series using Generalized Exponential Models. Journal of Statistical Planning and Inference, 139(6), 1992-2009.

Jansson, M. and Nielsen, M. O., (2012). Nearly Efficient Likelihood Ratio Tests of the Unit Root Hypothesis. Econometrica, 80(5), 2321-2332.

Jones, R. H. (1980) Maximum Likelihood fitting of ARMA models to time series with missing observations. Technometrics, 22, 389-395.

Jones, M. C. and Henderson, S. (2007). Nonparametric Time Series Analysis Using Kernel Estimators. Journal of the Royal Statistical Society: Series C (Applied Statistics), 56(4), 507-524.

Kalman, R. E. (1961). A new approach to linear filtering and prediction problems. Transactions of the American Society of Mechanical Engineers, 83D, 35-45.

Kalman, R. E. and Bucy, R. S. (1961). New results in linear filtering and prediction theory. Transactions of the American Society of Mechanical Engineers, 83, 95-108.

Karakostas, G. S. and Tsionas, E. G. (2005). Nonparametric Estimation of Long-Memory Processes. Journal of Time Series Analysis, 26(1), 1-16.

Koopman, S. J., Ooms, M. and Carnero, M. A. (2007). Periodic Seasonal reg-arfima-garch models for daily electricity spot prices. Journal of the American Statistical Association, 102(477), 16-27.

Lee, D. and Schmidt, P. (1996). On the Power of the KPSS Test of Stationarity against Fractionally Integrated Alternatives. Journal of Econometrics, 73, 285-302.

Leipus, R. and Viano, M. (2000). Modeling Long-memory Time Series with Finite or Inf infinite Variance: a General Approach. Journal of Time Series Analysis, 21(1), 61-74.

Ling, S. and Li, W. K. (1997) On Fractionally Integrated Autoregressive Moving-Average Time Series Models with Conditional Heteroscedasticity. Journal of the American Statistical Association, 92(439), 1184-1194.

Lo, A. W. (1991). Long-Term Memory in Stock Market Prices. Econometrica, 59(5), 1279-1313.

Lobato, I. N. and Savin, N. E. (1998). Real and spurious long-memory properties of stock market data. Journal of Business and Economic Statistics, 16, 261-283.

Malfait, M. and Roose, D. (1997). Wavelets for the Analysis of Fractal Time Series. Journal of Computational and Applied Mathematics, 82(1-2), 91-101.

Martínez, F., Charte, F., Frías, M.P. and Martínez-Rodríguez, A.M., 2022. Strategies for time series forecasting with generalized regression neural networks. Neurocomputing, 491, pp.509-521.

McAleer, M. and Medeiros, M. C. (2008). A Multiple regime smooth transition Heterogeneous Autoregressive model for long memory and asymmetries. Journal of Econometrics, 147(1), 104-119.

Montanari, A., Rosso, R. and Taqqu, M. S. (2000). A seasonal fractional ARIMA model applied to Nile River monthly flows at Aswan. Water Resources Research, 36, 1249-1259.

Morris, M. J. (1977). Forecasting the Sunspot Cycle. Journal of the Royal Statistical Society, 140(4), 437-468.

Müller, H.-G. and Stadtmüller, U. (2005). Nonparametric Curve Estimation: From Steady-State to Adaptive Methods. Statistical Science, 20(4), 468-489.

Nielsen, M. Ø. and Frederiksen, P. H. (2005). Finite sample comparison of parametric, semiparametric, and wavelet estimators of fractional integration. Econometric Reviews, 24(4), pp.405-443.

Ohanissian, A., Russell, J. R. and Tsay, R. S. (2008). True or Spurious Long Memory? A NewTest. Journal of Business and Economic Statistics, 26(2), 161-175.

Ooms, M. (1995). Flexible seasonal long memory and economic time series. Technical Report EI-9515/A. Econometric Institute, Erasmus University, Rotterdam.

Palma, W. (2007). Long Memory Time Series Theory and Methods. John Wiley and Sons, New Jersey.

Palma, W. and Chan, N. H. (2005). Efficient Estimation of Seasonal Long-Range-Dependent Processes. Journal of Time Series Analysis, 26(6), 863-892.

Pearlman, J. G., (1980). AnAlgorithmfortheExactLikelihoodofahigh-orderautoregressivemoving average process. Biometrika, 67(1), 232-233.

Peiris, M. S. (2003) Improving the Quality of Forecasting using Generalized AR models: An application to Statistical Quality Control. Statistical Methods, 5(2), 156-171.

Peiris, S., Allen, D. and Peiris, U. Generalized Autoregressive Models with Conditional Heteroscedasticity: an application to financial time series modeling. Proceedings of the Workshop on Research Methods: Statistics and Finance, 75-83, 2005.

Peiris, S. and Thavaneswaran, A. (2007) An introduction to volatility models with indices. Appl. Math. Lett., 20, 177-182.

Percival, D. B., and Walden, A. T. (2000). Wavelet methods for time series analysis. Cambridge University Press. https://doi.org/10.1017/CBO9780511841040

Phillips, P. C. B. and Xiao, Z. (1998). A Primer on Unit Root Testing. Journal of Economic Surveys, 12(5), 423-470.

Polya, G. and Szego, G. (1992). Problems and Theorems in Analysis I. Springer, New York.

Porter-Hudak, S. (1990). An application of the seasonal fractionally differenced model to the monetary aggregates. Journal of the American Statistical Association, Applic. Case Studies, 85, 338-344.

Rangarajan, G. and Ding, M, editors. (2003). Processes with Long-Range Correlations. Springer, Berlin.

Ray, B. K. (1993). Modeling long-memory processes for optimal long-range prediction. Journal of Time Series Analysis, 14, 511-525.

Reisen, V. A. (1994). Estimation of the fractional difference parameter in the ARIMA(p,d,q) model using the smoothed periodogram. Journal of Time Series Analysis, 15, 335-350.

Reisen, V., Abraham, B. and Lopes, S. (2001). Estimation of parameters in ARFIMA processes: A simulation study. Communications in Statistics- Simulation and Computation, 30(4), 787-803.

Reisen, V. A., Rodrigues, A. L. and Palma, W. (2006). Estimation of Seasonal fractionally integrated processes. Computational Statistics and Data Analysis, 50(2), 568-582.

Robinson, P. M. (1991). Testing for strong serial correlation and dynamic conditional heteroskedasticity in multiple regression. Journal of Econometrics, 47, 67-84.

Robinson, P. M. (1994). Efficient Tests of Non-stationary Hypotheses. Journal of the American Statistical Association 89, 1420-1457.

Robinson, P. M. (1995). Log-periodogram regression of time series with long range dependence. The Annals of Statistics 23, 1048-1072.

Robinson, P. M. (2003). Long Memory Time Series, Oxford University Press, Oxford.

Rosenblatt, M. (1956). Remarks on Some Nonparametric Estimates of a Density Function. The Annals of Mathematical Statistics, 27(3), 832-837.

Rossi, E. and Santucci de Magistris, P. (2014). Estimation of Long Memory in Integrated Variance. Econometric Reviews 33(7), 785-814.

Shephard, N. (1996). Statistical aspects of ARCH and stochastic volatility. In D. R. Cox, D. B. Hinkley, and O. E. Barndorff-Nielsen, editors, Time Series Models: In Econometrics, Finance and Other Fields. Chapman Hall, London.

Shitan, M. and Peiris, S. (2008). Generalised Autoregressive (GAR) model: a comparison of maximum likelihood and Whittle estimation procedures using a simulation study. Commun. Statist. Simul. Comput., 37(3), 560-570.

Shitan, M. and Peiris, S. (2013). Approximate Asymptotic Variance-Covariance Matrix for the Whittle Estimators of GAR(1) Parameters. Communications in Statistics-Theory and Methods, 42(5), 756-770.

Shumway, R. H. and Stoffer, D. S. (2010). Time Series Analysis and It's Applications with R Examples. Springer Science + Business Media, New York.

Silverman, B. W. (1986). Density Estimation for Statistics and Data Analysis, CRC Press.

Simpson, D. G. and Taqqu, M. S. (2006). Long-Memory Processes and Fractional Differencing. The Annals of Mathematical Statistics, 27(1), 1-20.

Sowell, F. (1992). Maximum likelihood estimation of stationary univariate fractionally integrated time series models. Journal of Econometrics, 53, 165-188.

Taylor, A. M. R. (2005). Fluctuation Tests for a Change in Persistence. Oxford Bulletin of Economics and Statistics, 67(2), 207-230.

Teyssière, G. and Davis, R. A. (2004). A Nonparametric Estimation of the Hurst Parameter in Long-Memory Time Series. Journal of Econometrics, 122(2), 187-215.

Teyssiere, G. and Kirman, A., editors. (2007). Long Memory in Economics. Springer, Berlin.

Tong, H. (1990). Non-Linear Time Series: A Dynamical System Approach, Oxford: Oxford University Press.

Velasco, C. and Robinson, P. M. (2000). Whittle pseudo-maximum likelihood estimation for nonstationary time series. Journal of the American Statistical Association, 95, 1229-1243.

Wand, M. P. and Jones, M. C. (1995). Kernel Smoothing, CRC Press.

Wang, Q., Lin, Y. and Gulati, C. M. (2003). Asymptotics for General Fractionally Integrated Processes with Applications to Unit Root Tests. Econometric Theory, 19(1), 143-164.

Wang, X., Smith, K. and Hyndman, R. J. (2006). Characteristic-Based Clustering for Time Series Data. Data Mining and Knowledge Discovery, 13, 335-364.

Whittle, P. (1951). Hypothesis Testing in Time Series. Uppsala: Almqvist and Wiksells.

Wood, S. N. (2006). Generalized Additive Models: An Introduction with R, CRC Press.

Woodward, W. A., Chen, Q. C. and Gray, H. L. (1998). A k-factor GARMA long-memory model. Journal of Time Series Analysis, 19, 485-504.

Xu, Y. and Chen, X. (2019). Long-Memory Time Series Modeling Using Deep Learning. Journal of Machine Learning Research, 20(1), 1-30.

Yajima, Y. (1985). On Estimation of Long-Memory Time Series Models. Australian and New Zealand Journal of Statistics, 27(3), 303-320.

Zhang, G. P. (2003). Time Series forecasting using a hybrid ARIMA and neural network model. Neurocomputing, 50, 159-175.

Zivot, E. and Wang, J. (2003). Modeling Financial Time Series with S-PLUS, CRC Springer.

R Software Codes with and/or without Machine Learning

1. R Computer Program Source Code used to create realizations, acf, pacf, sdf of GARMA(0,d,0) series driven by white noise errors.

```
> set.seed(1)
> z = rnorm(500)
> coeff = numeric(499)
> t = numeric(500)
> y = numeric(500)
> for(nin3 : 499)+l = 0.1+x = 0.8+coeff[1] = 1+coeff[2] = 0.16+coeff[n] =
(((2 * x) * (n + l - 1) * coeff[n - 1]) - ((n + (2 * l) - 2) * coeff[n - 2]))/n
> t = c(1, coeff[1 : 499])
> M = NULL
> N = length(t)
> for(iin1 : N) + M = rbind(M, c(t[i : 1], rep(0, N - i)))
> y = M
> ts.plot(y)
> acf = acf(y)
> pacf = acf(y, type = "partial")
> omega = -314 : 314
> omega = omega/100
> u = 0.8
> sigma2 = (sd(z)2)
> fxw = (sigma2(2pi))((4(cos(omega)u)2)(1))
> plot(omega, fxw,' l')
```

2. R Computer Program Source Code used to create realizations, acf, pacf and sdf of GARMA(0,d,0)-GARCH(1,1) series.

```
> set.seed(1)
> require(TSA)
> t = numeric(400)
> x = garch.sim(alpha = c(0.4, 0.3), beta = c(0.3), n = 1000)
> y = x[201 : 1000]
> p = numeric(400)
> coeff = numeric(800)
> g = numeric(400)
> for(jin3 : 800) + l = 0.45 + v = 0.8 + coeff[1] = 1 + coeff[2] = 0.72 +
coeff[j] = (((2 * v) * (j + l - 1) * coeff[j - 1]) - ((j + (2 * l) - 2) * coeff[j - 2]))/j
> cone = coeff[1 : 400]
> ctwo = coeff[401 : 800]
> k = 1 : 400
```

```
> for(i in 401 : 800) + g[i] = sum(cone[k] * y[i - k])
> t = g[401 : 800]
> for(n in 1 : 400) + p[n] = sum(cone[k] * ctwo[n])
> Realization = ts.plot(t)
> sampleacf = acf(t)
> Trueacf = acf(p)
> omega = -314 : 314
> omega = omega/100
> sigma2 = (sd(y)2)
> fxw = (sigma2((4 * cos(omega)v)2)(045))
> spectrum = plot(omega, fxw, "l")
> postscript(file = Colomboplot1epswidth = 12height = 7)
> par(mfrow = c(3, 2))
> ts.plot(t)
> acf(t)
> acf(p)
> acf(t, type = "partial")
> plot(omega, fxw, "l")
> dev.off()
```

3. An R program to assess the forecast accuracy of a GRNN model in ML with the "rolling_origin" function.

```
> pred < -grnn_forecasting(ts(1 : 20), h = 4, lags = 1 : 2)
> ro < -rolling_origin(pred, h = 4)
> print(ro$test_sets)
> print(ro$predictions)
> print(ro$errors)
> ro$global_accu
> ro$h_accu
> plot(ro, h = 4)
```

Answers to Selected Chapter Exercises

Chapter 1.

1. Yes. Pacf does corroborate the acf.

2. Trend, Seasonality, Cycles, and Random (irregular) variations or noise.

3. $X(t) = T(t) + C(t) + S(t) + I(t)$ and $X(t) = T(t).C(t).S(t).I(t)$

7. Method of moments, Maximum likelihood, and Ordinary Least Squares estimation methods.

9. In an AR(1) process generally represented as $X_t = \phi_1 X_{t-1} + w_t$, when $\phi_1 = 1$ the process becomes equivalent to a random walk.

Chapter 2.

1. $1 - \sum_{i=1}^{p} \phi_i B^i$ and $1 + \sum_{i=1}^{q} \theta_i B^i$.

2. ARMA model becomes a digital filter with white noise.

3. An exponential decay, since it's a short memory model.

4. An exponential decay, since it's a short memory model.

5. AIC and BIC.

Chapter 3.

1. $(1 - \sum_{i=1}^{p} \phi_i L^i)$ and $(1 + \sum_{i=1}^{q} \theta_i L^i)$

3. Exponential decay.

4. Exponential decay.

11. Exponential decay with seasonal spikes.

12. Exponential decay with seasonal spikes.

Chapter 4.

1. In the differencing equation $X_t = (I - B)^\delta Y_t$, when $\delta \in (0, 0.5)$ with δ being equal to fractional values, it results in fractional differencing. If δ is equal to an integer with $\delta \in (-0.5, 0.5)$, then it illustrates integer differencing.

2. ARFIMA model is a fractionally differenced (or fractionally integrated) model, and the ARIMA model does not constitute of a fractional differencing filter.

3. Integer differencing of a short memory ARMA model provides an ARIMA model. Fractional differencing of an ARIMA model results in a long memory ARFIMA model if the fractional differencing parameter d satisfies $d \in (0, 0.5)$.

6. The decaying ACF and PACF plots illustrate the memory type of a fractionally differenced process.

7. An exponential decay of ACF and PACF would depict a fractionally differenced short memory model as opposed to a hyperbolic decay that shall illustrate a fractionally differenced long memory model. Any ACF and PACF plots with shapes in between exponential and hyperbolically decaying curves could be defined as fractionally differenced intermediate memory models.

Chapter 5.

2. Autocorrelation, partial autocorrelation, and spectral density functions.

5. State Space modeling.

6. Linearity and Gaussianity features of a time series.

7. A set of recursions used to filter noise.

10. Both a standard long memory process and a Gegenbauer generalized long memory process would depict hyperbolic decay of ACF and PACF functions. However, they could be differentiated and distinguished by visually inspecting the spectral density function (sdf). In standard long memory models the sdf will have infinite peaks close to the origin, but on the contrary in generalized long memory models, the infinite peaks of the sdf will be away from the origin.

Chapter 6.

1. Standard long memory.

2. Hyperbolically decaying ACF and PACF functions and an SDF plot with infinite peaks close to the origin.

3. Measurement (Observation) equation and the Transition (State) equation.

6. MA approximation and AR approximation.

7. It is feasible to use quasi-maximum likelihood as opposed to maximum likelihood in estimating and predicting time series parameters, since the underlying statistical distribution of the time series process is unknown and the utilization of an approximate likelihood would result in a minimal margin of error. Furthermore, using quasi-maximum likelihood would enable the linearizing of a time series less cumbersome through the deployment of the Wald expansion.

Chapter 7.

1. Generalized long memory.

2. Hyperbolically decaying ACF and PACF functions and an SDF plot with infinite peaks away from the origin.

3. Measurement (Observation) equation and the Transition (State) equation. But they will be slightly different from the corresponding equations linked with an ARFIMA model.

5. Linearity and Gaussianity of the GARMA time series.

6. MA approximation and AR approximation. They are cast in state space form through linearization done employing the Wald expansion.

Chapter 8.

1. If unity (or 1) is the root of a characteristic equation that governs a time series, then such a time series is defined as a linear time series.

2. If unity (or 1) is not a root of a characteristic equation that governs a time series, then such a time series is defined as a nonlinear time series.

3. If the roots of a characteristic equation lie inside the unit circle with a modulus or absolute value less than 1, then the first difference of the process will be defined as a stationary time series.

4. If the roots of a characteristic equation do not lie inside the unit circle with a modulus or absolute value less than 1 and if it does possess a unit root of 1, then such a process is non-stationary.

5. A unit root is the root of a characteristic equation of a time series process that is equal to 1.

Chapter 9.

All the questions in this section are based on running R programs using simulated and downloaded data. Therefore, precise answers cannot be provided.

Chapter 10.

1. White noise is independent and identically distributed (i.i.d) errors, whereas GARCH errors are white noise terms based on history that are not i.i.d.

7. Unconditional heteroskedasticity refers to general structural changes in volatility that are not related to prior period volatility. Conditional heteroskedasticity identifies nonconstant volatility related to prior period's volatility. Volatility is the characteristic of changing often and unpredictably.

Chapter 11.

1. Machine learning (ML) is the use and development of computer systems that are able to learn and adapt without following explicit instructions, by using algorithms and statistical models to analyze and draw inferences from patterns in data.

2. High-frequency data refer to time series data collected at a very fine scale, often at intervals of seconds or even milliseconds.

3. Supervised learning, unsupervised learning and reinforced learning.

4. tsfGRNN package is an unsupervised learning technique in ML created by combining time series forecasting methodologies with generalized regression neural networks (GRNNs) in ML.

11. Seasonality.

12. Seasonality and long memory.

13. Disease incidence frequency varies with time.

14. Fractional integration.

Index

Note: Page numbers in **bold** and *italics* refer to tables and figures, respectively.

For Product Safety Concerns and Information please contact our EU
representative GPSR@taylorandfrancis.com
Taylor & Francis Verlag GmbH, Kaufingerstraße 24, 80331 München, Germany

www.ingramcontent.com/pod-product-compliance
Lightning Source LLC
Chambersburg PA
CBHW070726220326
41598CB00024BA/3316